Piezoelectric Actuators

Edited by Tinghai Cheng and Jianping Li

Published in London, United Kingdom

Piezoelectric Actuators
http://dx.doi.org/10.5772/intechopen.97944
Edited by Tinghai Cheng and Jianping Li

Contributors
Tadej Rojac, Jinguo Han, Mingming Lu, Jieqiong Lin, Min Wang, Xuan Fang, Songquan Liao, Shibo Fu, Matej Možek, Borut Pečar, Danilo Vrtačnik, Drago Resnik, Jianping Li, Shupeng Wang, Shitong Yang, Yuelong Li, Guangda Qiao, Xiaosong Zhang, Xiaohui Lu, Tinghai Cheng, Keping Liu, Zhenguo Zhang, Piao Fan, Yikun Dong, Shuai Yu

First published in London, United Kingdom, 2022 by IntechOpen
IntechOpen is the global imprint of INTECHOPEN LIMITED, registered in England and Wales, registration number: 11086078, 5 Princes Gate Court, London, SW7 2QJ, United Kingdom

British Library Cataloguing-in-Publication Data
A catalogue record for this book is available from the British Library

Additional hard and PDF copies can be obtained from orders@intechopen.com

Piezoelectric Actuators
Edited by Tinghai Cheng and Jianping Li
p. cm.
Print ISBN 978-1-80355-915-5
Online ISBN 978-1-80355-916-2
eBook (PDF) ISBN 978-1-80355-917-9

Meet the editors

Dr. Tinghai Cheng received a B.S., M.S., and Ph.D. from Harbin Institute of Technology, China, in 2006, 2008, and 2013, respectively. He was a visiting scholar in the School of Materials Science and Engineering, Georgia Institute of Technology, USA, under the supervision of Prof. Zhong Lin Wang from 2017 to 2018. Currently, he is a professor at the Beijing Institute of Nanoenergy and Nanosystems, Chinese Academy of Sciences, and Changchun University of Technology. As the first or corresponding author/co-author, Dr. Cheng has to date published seventy-three SCI academic articles. He also has thirty-six patents and eleven software copyrights to his credit.

Dr. Jianping Li received a B.S. and Ph.D. from the School of Mechanical Science and Engineering, Jilin University, Changchun, China, in 2011 and 2016, respectively. He worked with Chiba University, Japan, as a JSPS Researcher, supported by the Japan Society for the Promotion of Science (JSPS). Since 2018, he has been working with the Institute of Precision Machinery and Smart Structure, College of Engineering, Zhejiang Normal University, China. To date, he has published more than seventy academic papers. He also has one international and fourteen Chinese patents to his credit, and he has participated in the formulation of two industry standards.

Contents

Preface

Demand for manufacturing has increased with the development and progress of science and technology. In this regard, piezoelectric actuators and transducers are attracting attention because of characteristics such as millisecond response speed, nanoscale accuracy, power-off self-locking, absence of electromagnetic interference, and no generation of a magnetic field. Owing to their advantages over traditional electromagnetic actuators, piezoelectric actuators and transducers have been used in aerospace engineering, biomedical engineering, artificial intelligence, micro-nano processing, 3D printing, and more. Researchers have developed many novel types of precision piezoelectric actuators and transducers to improve output performance, including velocity, load capacity, and accuracy. This book comprehensively summarizes the latest progress in precision piezoelectric actuators and transducers, such as piezoelectric bionic actuators, piezoelectric stick-slip actuators, piezoelectric direct-driving actuators, and piezoelectric pumps. It also discusses corresponding working principles, excitation signals, design theoretical framework, novel structure, and control methods. This book is a useful reference that provides guidance for the development of piezoelectric actuators and transducers.

Chapter 1 introduces piezoelectric actuators based on different operating principles, including direct-driving piezoelectric actuators, ultrasonic piezoelectric actuators, friction-inertia piezoelectric actuators, and bionic piezoelectric actuators.

Chapter 2 presents an intriguing case where domain walls (DWs) exhibit enhanced electrical conductivity with respect to bulk conductivity. By combining experimental data and modeling, it will be shown that local conductivity, related to the accumulation of charged point defects at DWs, not only affects DW dynamics through DW-defect pinning interactions but also affects the macroscopic nonlinearity and hysteresis in a more complex manner. The chapter also reviews and discusses the major characteristics and implications of the underlying nonlinear Maxwell–Wagner piezoelectric relaxation, triggered by the presence and dynamics of conducting DWs, in the framework of systematic multiscale analyses on BiFeO3 ceramics.

Chapter 3 introduces bionic-type piezoelectric actuators with long-range outputs. First, the chapter discusses commonly used piezoelectric materials and piezoelectric effects, including positive piezoelectric effect and inverse piezoelectric effect. Second, it elaborates on inchworm-type actuators and seal-type actuators, which are classified into the walker type, the pusher type, and the mixed type. Finally, it examines the characteristics of bionic actuators, such as their configurations, classifications, principles, connections, and distinctions.

Chapter 4 introduces a piezoelectric tool actuator used in elliptical vibration cutting, which offers tertiary cutting operations with quick response and flexible modulated ability. The chapter covers the working principles of piezoelectric tool actuators, compliant mechanism design, static modeling, kinematic and dynamic modeling, structure optimization, and offline testing.

Chapter 5 briefly presents three distinctive concepts of micropump actuator driving modules, each with its waveform specifics and impact on micropump performance. The first presented concept is based on two mutually exclusive boost switched-mode power supply modules. Characterization of this module identified output voltage asymmetry to be the limiting factor of micropump performance. To assure driving symmetry, an alternative driving module based on independent high-voltage stages and optocouplers was implemented. Moreover, the design is based on an embedded arbitrary waveform generator, which offers an efficient trade-off between high pumping performance and low current consumption.

Chapter 6 introduces the development of topology optimization and summarizes the design and research of the compliant mechanism of the piezoelectric actuator. The experiments show that the topology optimization method has guiding significance for the design of piezoelectric actuators. The chapter ends by proposing future research directions and challenges of topology optimization design for flexure hinge-type piezoelectric actuators.

Chapter 7 studies the active vibration suppression of an active-passive composite vibration suppression system based on piezoelectric actuators. On the basis of fully analyzing the characteristics of the piezoelectric actuator and displacement amplifying mechanism and the dynamic model of the vibration suppression system, an active composite control strategy based on IFF feedback and RLS adaptive feedforward for vibration suppression on the piezoelectric system is discussed.

Chapter 8 introduces the modeling and control of piezoelectric stick-slip actuators. In the aspect of modeling, the existing mathematical models describing the hysteresis characteristics of piezoelectric stick-slip actuators and the mathematical models of complex friction relationships in the structure are introduced. In terms of control, according to open-loop control and closed-loop control, this chapter summarizes and studies the efforts made to make up for control accuracy, and summarizes many control cases, such as feedforward control, sliding mode control, PID control, neural network control, and others.

We are very grateful for the help of our colleagues who contributed to this book.

Tinghai Cheng
Beijing Institute of Nanoenergy and Nanosystems,
Chinese Academy of Sciences,
Beijing, China

Jianping Li
College of Engineering,
Zhejiang Normal University,
Jinhua, China

Section 1

Introduction

Chapter 1

Introductory Chapter: Piezoelectric Actuators

Tinghai Cheng and Jianping Li

1. Introduction

Micro/nano-positioning system with micro-/nano-accuracy is a key technology in industry and science fields such as precision machining and measurement, optical engineering, modern medical treatment, biological genetic engineering, and aerospace science and technology [1–4]. The so-called actuators refer to a functional device that can output specific motion, such as linear and rotary motion, for micro-/nano-positioning systems. Traditional actuators, such as electrical motors, hydraulic motors, and pneumatic motors, can realize the large stroke and large load; however, their positioning accuracy is low and the size is large. They are still facing the problem of high-positioning resolution and compact size. In recent decades, researchers

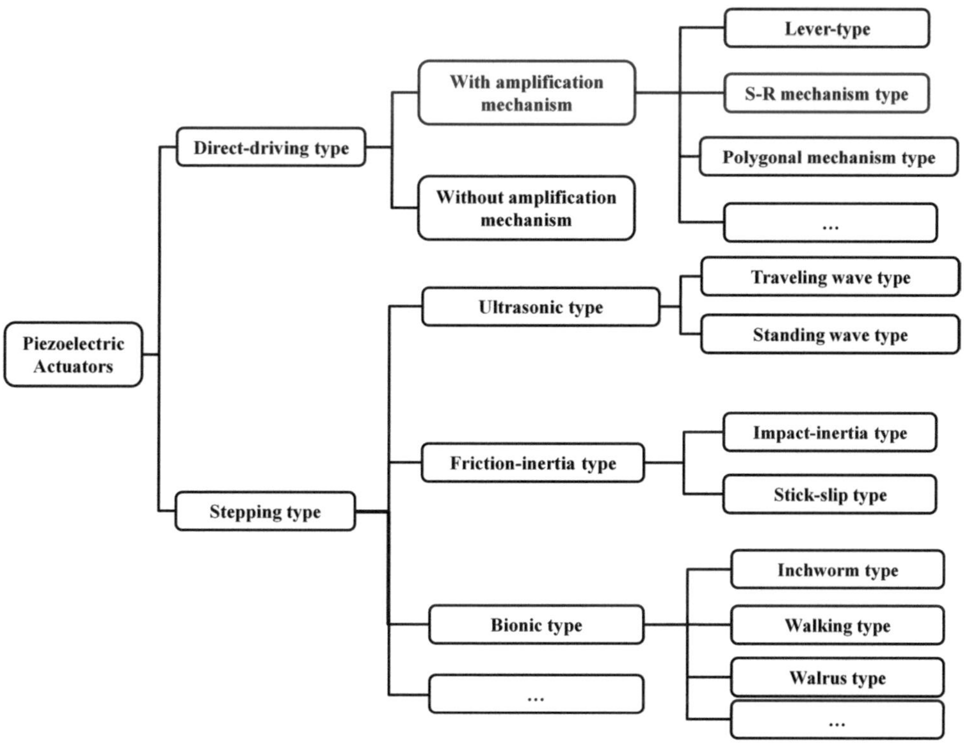

Figure 1.
Classification of piezoelectric actuators.

devote themselves to the research of new actuators with better performance to realize micro-/nano-accuracy with a compact size. Piezoelectric actuator is one novel actuator that uses the inverse piezoelectric effect of piezoelectric materials to convert electrical energy into mechanical energy to realize controllable positioning accuracy [5, 6]. Piezoelectric actuators have the characteristics of compact size, light weight, high precision, fast response, good control characteristics, high energy density, low energy consumption, and free from magnetic field interference. Researchers have developed a variety of piezoelectric actuators and applied them to biological cell micromanipulation, atomic manipulation, micro-/nano-indentation and other systems, and achieved good application results. According to different driving principles, generally, the piezoelectric actuators can be divided into two categories: direct-driving piezoelectric actuators and stepping piezoelectric actuators. The direct-driving piezoelectric actuators mainly apply piezoelectric elements to directly drive the output mechanism, and its working stroke is usually small; the stepping piezoelectric actuators adopt the stepping motion mode to realize large working displacement, which can be further subdivided into ultrasonic piezoelectric actuator, friction-inertia piezoelectric actuator, bionic piezoelectric actuator, etc., as shown in **Figure 1**.

2. Overview of piezoelectric actuators

Direct-driving piezoelectric actuators mainly utilize piezoelectric elements (such as piezoelectric stack, piezoelectric bimorph) to directly push the output mechanism, which could be easy to achieve significant advantages of high output accuracy, large output load, and compact size structure [7, 8]. The development of flexure hinge technology has greatly expanded the application field of piezoelectric actuators. Since the end of last century, researchers from the United States, Japan, Australia, Germany, China, and other countries have competed to develop various types of direct-driving piezoelectric actuators based on flexure hinge mechanism, which can be further divided into piezoelectric actuators without amplification mechanism and piezoelectric actuators with amplification mechanism.

Ultrasonic piezoelectric actuator, also known as ultrasonic motor, is one stepping piezoelectric actuator developed rapidly in the 1980s [9, 10]. It is based on inverse piezoelectric effect and the principle of ultrasonic vibration. In the working process, the micro-elliptical resonance of the elastomer in the ultrasonic frequency range is excited by the inverse piezoelectric effect of piezoelectric materials and transformed into the rotation or linear motion of the mover through the action of friction, so as to achieve the required output displacement and load. According to different wave propagations, they can be divided into standing wave ultrasonic piezoelectric actuators and traveling wave ultrasonic piezoelectric actuators. Commercial applications have been obtained for ultrasonic piezoelectric actuators.

The motion principle of friction-inertia piezoelectric actuator is based on the law of conservation of momentum [5, 6, 11]. By applying voltage signals, such as sawtooth wave to piezoelectric elements, the relative displacement between stator and mover or two mass blocks with different mass is generated, so as to realize step-by-step large stroke motion. The structure of friction-inertia piezoelectric actuator is relatively simple, and sometimes, the displacement output with high accuracy can be realized only with two moving parts (stator and mover), and the control strategy is also relatively simple. Therefore, it has attracted the continuous attention of researchers all over the world. According to different motion principles, it can be further divided into impact-inertia piezoelectric actuator and stick-slip piezoelectric actuator.

Bionic piezoelectric actuator is one novel piezoelectric actuator, which mimics the motion style of different creatures in the nature to overcome the limitation of traditional piezoelectric actuators [2, 12, 13]. Bionic piezoelectric actuators are able to achieve large working stroke or large output force, which is of great significance for the development of piezoelectric actuators. Based on different motions of creatures, bionic piezoelectric actuator can be further divided into inchworm-type piezoelectric actuators, walking-type piezoelectric actuators, walrus-type piezoelectric actuators, and so on.

Up to now, many kinds of piezoelectric actuators have been proposed and investigated, and some kinds of piezoelectric actuators have been applied into real applications. However, due to the friction and wear of materials, how to achieve the high accuracy and long-term reliability for piezoelectric actuators is still a problem. This book introduces some basic foundations of piezoelectric actuators, including the piezoelectric phenomenon, the modeling and control of piezoelectric actuators, different kinds of piezoelectric actuators, and some applications in different fields of piezoelectric actuators. We hope this book could give an overview of piezoelectric actuators for the new researchers to get a basic introduction.

Author details

Tinghai Cheng[1,2]* and Jianping Li[3]

1 School of Mechatronic Engineering, Changchun University of Technology, Changchun, China

2 Beijing Institute of Nanoenergy and Nanosystems, Chinese Academy of Sciences, Beijing, China

3 Key Laboratory of Intelligent Operation and Maintenance Technology and Equipment for Urban Rail Transit of Zhejiang Province, College of Engineering, Zhejiang Normal University, Jinhua, China

*Address all correspondence to: chengtinghai@binn.cas.cn

References

[1] Qiao G, Li H, Lu X, Wen J, Cheng T. Piezoelectric stick-slip actuators with flexure hinge mechanisms: A review. Journal of Intelligent Material Systems and Structures. 2022. DOI: 10.1177/1045389X211072244

[2] Li J, Huang H, Morita T. Stepping piezoelectric actuators with large working stroke for nano-positioning systems: A review. Sensors and Actuators A: Physical. 2019;**292**:39-51. DOI: 10.1016/j.sna.2019.04.006

[3] Li H, Liu J, Li K, Liu Y. A review of recent studies on piezoelectric pumps and their applications. Mechanical Systems and Signal Processing. 2021;**151**:107393. DOI: 10.1016/j.ymssp.2020.107393

[4] Salim M, Salim D, Chandran D, Aljibori HS, Kherbeet AS. Review of nano piezoelectric devices in biomedicine applications. Journal of Intelligent Material Systems and Structures. 2018;**29**:2105-2121. DOI: 10.1177/1045389X17754272

[5] Hunstig M. Piezoelectric inertia motors-a critical review of history, concepts, design, applications, and perspectives. Actuators. 2017;**6**:7. DOI: 10.3390/act6010007

[6] Cheng T, He M, Li H, Lu X, Zhao H, Gao H. A novel trapezoid-type stick-slip piezoelectric linear actuator using right circular flexure hinge mechanism. IEEE Transactions on Industrial Electronics. 2017;**64**:5545-5552. DOI: 10.1109/TIE.2017.2677318

[7] Pinskier J, Shirinzadeh B, Clark L, Qin Y. Development of a 4-DOF haptic micromanipulator utilizing a hybrid parallel-serial flexure mechanism. Mechatronics. 2018;**50**:55-68. DOI: 10.1016/j.mechatronics.2018.01.007

[8] Li Y, Xu Q. A totally decoupled piezo-driven XYZ flexure parallel micropositioning stage for Micro/Nanomanipulation. IEEE Transactions on Automation Science and Engineering. 2011;**8**:265-279. DOI: 10.1109/tase.2010.2077675

[9] Kurosawa MK, Kodaira O, Tsuchitoi Y, Higuchi T. Transducer for high speed and large thrust ultrasonic linear motor using two sandwich-type vibrators. IEEE Transactions on Ultrasonics, Ferroelectrics, and Frequency Control. 1998;**45**:1188-1195. DOI: 10.1016/j.sna.2020.112037

[10] Zhu Y, Yang T, Fang Z, Shiyang L, Cunyue L, Yang M. Contact modeling for control design of traveling wave ultrasonic motors. Sensors and Actuators A: Physical. 2020;**310**:112037. DOI: 10.1109/58.726442

[11] Qiao G, Ning P, Xia X, Yu Y, Lu X, Cheng T. Achieving smooth motion for piezoelectric stick–slip actuator with the inertial block structure. IEEE Transactions on Industrial Electronics. 2022;**69**:3948-3958. DOI: 10.1109/TIE.2021.3073314

[12] Wang R, Hu Y, Shen D, Ma J, Li J, Wen J. A novel piezoelectric inchworm actuator driven by one channel direct current signal. IEEE Transactions on Industrial Electronics. 2021;**68**:2015-2023. DOI: 10.1109/TIE.2020.2975493

[13] Li J, Zhao H, Qu X, Qu H, Zhou X, Fan Z, et al. Development of a compact 2-DOF precision piezoelectric positioning platform based on inchworm principle. Sensors and Actuators A: Physical. 2015;**222**:87-95. DOI: 10.1016/j.sna.2014.12.001

Section 2

Piezoelectric Properties

Chapter 2

Piezoelectric Nonlinearity and Hysteresis Arising from Dynamics of Electrically Conducting Domain Walls

Tadej Rojac

Abstract

Macroscopic nonlinearity and hysteresis observed in the piezoelectric and dielectric responses of ferroelectric materials to external stimuli are commonly attributed to localized displacements of domain walls (DWs). The link between the macroscopic response and microscopic DW dynamics is provided by the well-known Rayleigh relations, extensively used to quantify the electrical and electromechanical behavior of ferroelectric ceramics and thin films under subswitching conditions. In this chapter, I will present an intriguing case where DWs exhibit enhanced electrical conductivity with respect to the bulk conductivity. By combining experimental data and modeling, it will be shown that the local conductivity, related to accumulation of charged points defect at DWs, does not only affect DW dynamics through DW-defect pinning interactions, as we may expect, but goes beyond it by affecting the macroscopic nonlinearity and hysteresis in a more complex manner. The major characteristics and implications of the underlying nonlinear Maxwell-Wagner piezoelectric relaxation, triggered by the presence and dynamics of conducting DWs, will be presented, reviewed and discussed in the framework of systematic multiscale analyses on $BiFeO_3$ ceramics. The result may have implications in the development of promising $BiFeO_3$-based compositions for high-temperature piezoelectric applications.

Keywords: Piezoelectric, nonlinearity, hysteresis, domain wall, electrical conductivity, $BiFeO_3$

1. Introduction

Historically speaking, probably the most important examples of dielectric and piezoelectric nonlinearity and hysteresis in ferroelectrics can be found already in the initial lines of the introductory section of the paper by Joseph Valasek from 1921 [1], which we use today to celebrate the 100th anniversary of the discovery of ferroelectricity [2]. Despite struggling with moisture-sensitive Rochelle salt crystals, Valasek finally managed to publish the first ever case of a ferroelectric hysteresis loop. Today, such hysteresis of the polarization response to the external electric field, originating from ferroelectric domain switching, is essential in, e.g., ferroelectric random-access memories (FeRAMs) [3]. Hysteretic and nonlinear responses to external fields, however, are not only observed at switching conditions. Deviations from simple linear relationship between polarization (P) and electric-field (E) (dielectric response), P and mechanical stress (Π) (direct piezoelectricity) and mechanical strain (x) and E (converse piezoelectricity), are very common at subswitching driving conditions at which many important devices operate, such as capacitors, sensors and actuators. Such responses are measured in both bulk and thin-film ferroelectrics [4–6].

The origins of subswitching macroscopic nonlinearity and hysteresis alongside the frequency dependence of the property coefficients, which are directly relevant for device performance, are difficult to assess due to complex interrelated mechanisms often operating simultaneously over a wide length-scale range of the material. By risking of being too general, one could say that the major players of these mechanisms are domain walls (DWs) and similar interfaces [7], charged point defects [8–11], grain boundaries [12], spontaneous ferroelastic strains [13, 14], oxygen octahedra tilts [15] and secondary phases [16], if only the widely discussed examples are listed. Most of these features are always present in polycrystalline ferroelectric materials and are capable of cross-interacting via electric and/or elastic fields (a classical example that every ferroelectrician should be aware of is the DW-defect electro-elastic interaction [17, 18]). Complicating the picture is the fact that many of these features are strongly dependent on the synthesis conditions. Despite playing a key role, some of them, such as point defects, could not be directly visualized until recently [19–22], meaning that the interpretations of their role in the macroscopic response are supported by either indirect analysis or other reasoning (see next section). Without extensive knowledge in the area and with just a little intuition, it is pretty easy to understand why dielectric and piezoelectric nonlinearity as well as hysteresis in ferroelectric and similar materials are so difficult to understand and predict. Based on my personal opinion, the results shown in this chapter are the perfect example to illustrate the difficulties in elucidating the mechanisms responsible for the piezoelectric nonlinearity and hysteresis in a complex material, such as $BiFeO_3$ (BFO). The results that I am going to present have been acquired over eight long years of multidisciplinary research focusing on theoretical and experimental studies and involving a large number of researchers from different fields.

One of the most important and widely-recognized origins of nonlinearity and hysteresis in the subswitching electrical and electromechanical response of ferroelectrics is the displacement of domain walls (DWs) [5, 23–26]. These are interfaces separating domain regions in which the spontaneous polarization is directed in one of the symmetry-allowed directions. Displacements of DWs are mechanistically treated via the interaction with charged points defects, which act as pinning sites and, depending on their type, mobility and location, affect the movement of DWs under applied fields (this issue will be thoroughly discussed in the next section). Reversible and irreversible motion of DWs in a ferroelectric with randomly distributed pinning sites (defects) is described by the Rayleigh law (RL), which is vastly used to quantify and cross-compare the subswitching responses of ferroelectric ceramics and thin films [6, 27]. For the converse piezoelectric effect (same is valid for the direct effect or dielectric response), the RL can be expressed by a set of two equations:

$$d'(E_0) = d_{init} + \alpha \cdot E_0 \tag{1}$$

$$x(E) = (d_{init} + \alpha \cdot E_0) \cdot E \pm \frac{\alpha}{2}\left(E_0^2 - E^2\right) \tag{2}$$

For simplicity, the subscripts of the piezoelectric coefficient are omitted. Eq. (1) describes the linear dependence of the real component of the piezoelectric coefficient (d') on the electric-field amplitude (E_0) with d_{init} and α representing the reversible and irreversible Rayleigh coefficients, respectively. Note that d_{init} corresponds to the coefficient extrapolated at zero field or $d_{init} = d'(E_0 = 0)$. Eq. (2) is a multivalued function and describes the strain (x) versus electric-field (E) relationship where the positive and negative signs, separating the first and the second term of the equation, are related to the strain branches defined by descending and ascending E, respectively. The second quadratic term of Eq. (2) thus defines the hysteresis. Note that *nonlinearity* refers here to the *nonlinear* relationship between strain (x) and electric field (E) (second quadratic term of Eq. (2)), in which case the proportionality coefficient, *d'*, is *linearly* dependent on the field amplitude (Eq. (1)).

By carefully inspecting the Rayleigh equations, it is not difficult to understand that the essence of this model is the intimate relationship between the nonlinearity and hysteresis. For example, in Eq. (1) the irreversible coefficient α defines the nonlinearity by the field dependence of the piezoelectric coefficient, however, α also appears in the second term of Eq. (2) describing the hysteresis (by the different up-field and down-field strain branches). Alternatively, one can easily realize the nonlinearity-hysteresis relation of RL in the second term of Eq. (2), which sets the nonlinear (quadratic) relation of strain to field while, at the same time, representing the hysteresis. In other words, hysteresis arises from nonlinearity and vice versa. A mathematical test is straightforward: in the absence of the irreversible contribution ($\alpha = 0$), there are two important consequences: (i) the coefficient *d'* becomes independent on E_0 (Eq. (1) reduced to $d'(E_0) = d_{init}$), meaning that the response is linear (i.e., the coefficient is constant), and (ii) the hysteresis vanishes (Eq. (2) reduces to the first anhysteretic term). Therefore, every irreversible DW displacement contributes to nonlinearity and hysteresis in a unique way set by the RL [27].

When mechanistically reasonable, the irreversible coefficient α, experimentally extracted from the measurements of, e.g., longitudinal converse piezoelectric coefficient d_{33} as a function of field amplitude E_0 (se Eq. (1)), is commonly used to quantify (irreversible) DW contribution to piezoelectric properties (an analogous approach is used to quantify DW contribution to dielectric permittivity). A rigorous test to find out whether the measured data can be approximated by RL is to at least validate the nonlinearity-hysteresis relationship implied by RL. This can be done by verifying whether the experimental hysteresis can be fitted by Eq. (2) using the coefficient α that is determined from the d_{33} versus E_0 slope [6, 27]. Importantly, other parameters of the measured response can be checked against the RL predictions. For example, using the Fourier-series analysis of Rayleigh equations (derivations can be found in Ref. [27]) and the hysteresis-area analysis in complex mathematical formalism (see, e.g., references [24, 28]), it can be shown that the ratio between the increment in the real piezoelectric coefficient ($\Delta d' = d'(E_0) - d_{init}$) and the increment of the imaginary piezoelectric coefficient ($\Delta d'' = d''(E_0) - d''(E_0 = 0)$) is constant (equal to $\frac{4}{3\pi}$) and does not depend on the external field amplitude. This can be easily verified if the experimental piezoelectric coefficients and piezoelectric phase angles (or dielectric permittivity and losses) are known for different driving fields [14]. Other response parameters that can be evaluated against the RL predictions are embedded in the third harmonic response; this is discussed in detail in Ref. [29].

It has been shown that the subswitching response of a number of ferroelectric materials well obey the RL. Examples include donor-doped soft $Pb(Zr,Ti)O_3$ (PZT) ceramics [5, 27, 28, 30], coarse-grained undoped $BaTiO_3$ [31], high-Curie-temperature piezoceramics based on $BiScO_3$-$PbTiO_3$ [24], textured $(K_{0.5}Na_{0.5})NbO_3$ (KNN) ceramics [32], Aurivilius-type Nb-doped $Bi_4Ti_3O_{12}$ [31], some relaxor-ferroelectric Pb $(Mg_{1/3}Nb_{2/3})O_3$-$PbTiO_3$ compositions [33, 34] and Pb-based thin films [35, 36]. It has to be pointed out, however, that even in these cases, strictly speaking, a good match between the Rayleigh model and the measured subswitching response is typically observed in a limited driving field range, sometimes referred to as the "Rayleigh range" [5, 27]. This is expected, considering that RL is an idealization and assumes DW motion in a perfectly random pinning potential. It is clear that real materials' responses may come close to this situation; however, in most cases deviations from RL predictions are naturally observed and should not be surprising. It is legitimate to think that these deviations have microscopic origins different from DW motion. However, this is not necessarily the case. A nice supporting example can be found in soft PZT, which often exhibit sublinear, quasi-saturating field dependency of the piezoelectric coefficient at relatively weak fields, clearly deviating from the perfect linear behavior predicted by Eq. (1) [4]. In this case, Preisach approach becomes useful as it shows that such response can be understood by considering a non-uniform pinning potential where the concentration of weakly pinned DWs is higher than those that are strongly pinned (note that the two should be equal in a perfect Rayleigh random pinning potential, resulting in a flat Preisach distribution function) [37]. Therefore, while the experimental data deviates from the RL, in some cases this deviation can still be explained by DW motion without necessarily invoking other mechanisms unrelated to DWs.

Apart from those cases where the response is close to RL predictions, the subswitching response can be clearly non-Rayleigh. Acceptor-doped hard PZT [30, 38], Sm-doped $PbTiO_3$ [27] or monoclinic $Pb(Mg_{1/3}Nb_{2/3})O_3$-$PbTiO_3$ (PMN-PT) ceramic compositions with relaxor features [34] are just few of such examples. In this chapter, I am going to present another case that has attracted particular attention in recent times. The case is on BFO, a perovskite that is unique for at least two reasons: (i) because it contains electrically conducting DWs that are formed spontaneously and are attractive for applications in nanoelectronics [39] and (ii) due its high Curie temperature (~835°C) [40, 41], which makes it the key perovskite in the development of novel compositions for high-temperature piezoelectric applications [42]. As it will be shown in the next section, enhanced conductivity at DWs completely alters the DW dynamics under applied subswitching fields. I will give evidence of a very large nonlinear contribution to the total piezoelectricity in BFO arising from irreversible DW dynamics, which is strongly dependent on the frequency of the external driving field. Along with a peculiar hysteretic behavior, characterized by a negative piezoelectric phase angle, and an unusual clock-wise rotational sense, I will strive to demonstrate how complex the dynamic response of electrically conducting DWs can be and how careful should we be in analyzing and treating materials that show unique behaviors. Somewhat surprisingly, I will show that despite the charged point defects, which are accumulated at DWs and are responsible for the local conductivity, act as pinning sites (as explained in Section 2), they can even increase the DW mobility and largely contribute to the enhanced nonlinear converse piezoelectric response. This may happen because the local conductive paths through the grains, set by the conductive DWs, lead to a redistribution of internal fields, effectively resulting in grains or grain families inside which the electric field is largely enhanced with respect to the field applied externally. The mechanism, called nonlinear Maxwell-Wagner piezoelectric effect (explained in Section 3), can be supported by simple analytical modeling (Section 4) and should be more seriously considered when engineering BFO-based materials for high-temperature piezoelectric applications.

2. Dynamic interactions between charged point defects and domain walls: a microscopic view

In this section, we will take a look at idealized microscopic pictures describing the interactions between DWs and charged point defects, which have profound implications in the DW dynamics and play a key role in the macroscopic dielectric and piezoelectric nonlinearity and hysteresis. Before discussing these interactions, however, it is timely to emphasize few important points. First, the reader should be aware that DW dynamics is not the only mechanism giving rise to electrical and electromechanical hysteresis. An example that should be familiar to every scientist and engineer working in the field of dielectrics or capacitors, is the DC electrical conductivity, which can directly contribute to the imaginary dielectric coefficient and thus dielectric losses and P-E hysteresis. In ionic compounds, electron and holes tend to be trapped at local sites in the lattice; they locally polarize and deform the lattice, creating the so-called polarons. Being self-trapped, these charges can move via an activated hopping process, similar to ionic conductivity [43]. Interestingly, in both physical and mathematical sense, hopping charge transitions are indistinguishable from pure dipole reorientations, meaning that hopping conductivity can contribute not only to losses (imaginary dielectric permittivity) but also to polarization (real dielectric permittivity). Mathematical descriptions of these problems are given in the book by Jonscher, which is highly recommended for learning conductivity phenomena in dielectrics [44].

Second, the effect of the electrical conductivity can go beyond a simple contribution to the complex dielectric permittivity and can strongly affect piezoelectricity, too. This case, which is less trivial and somewhat difficult to digest (even for experts in the field), is the Maxwell-Wagner (M-W) piezoelectric relaxation [45–49]. As an analogue of the well-known dielectric M-W relaxation [50], it originates, generally speaking, from electrically conductive paths present locally inside the material, which lead to internal electric-field redistribution as a function of the driving field frequency and, consequently, to electromechanical relaxation phenomena. The effect was predicted for the direct [4, 48] and converse piezoelectric response [51], and even for purely elastic response [52]. It was experimentally observed in heterogeneous ceramic materials [48], two-phase polymer systems [53] and piezoelectric-polymer composites [47]. The piezoelectric M-W effect is described in detail in an older book on hysteresis [4], to which the reader should refer to as a support to the data presented in this chapter where the focus is on BFO ceramics.

To explain the dielectric M-W effect, either grain boundaries or sample-electrode interfaces are commonly considered, as these are the regions where depletion layers are likely to be formed, giving rise to inhomogeneous field distribution inside the material and, consequently, to M-W relaxation [50]. In this chapter, I will present a different case where local conductive paths originate from charges concentrated at DWs. It is intuitively understood that this case must be unique because, unlike grain boundaries or sample-electrode interfaces, the conductive paths provided by the DWs are dynamic: they can move and locally displace under external fields. One may simplistically view this scenario as a local conductivity that is not spatially confined.

The third point to consider is rather general. While the discussion in this chapter is mostly about DWs, the reader should understand that also the dynamics of other types of interfaces may lead to the same macroscopic effects associated with subswitching nonlinearity and hysteresis. As an example, those other interfaces may be represented by boundaries separating regions consisting of phases with different crystallographic symmetries, which are present in morphotropic compositions [7]. The role of interface boundary motion during subcoercive loading of morphotropic $BiScO_3$-$PbTiO_3$ ceramics was recently demonstrated by in situ X-ray diffraction (XRD) analysis [54].

We come back now to our case on DWs. In analogy to the well-known pinning of DWs in ferromagnetic materials [55], the same concept is used for ferroelectrics [56]. An important difference is that, on average, ferroelectric DWs are an order of magnitude thinner than ferromagnetic DWs [39, 57–59]. While it is clear that the subject is complex and cannot be reduced to few (over)simplified pictures, like those shown in **Figure 1**, it is surprising to realize that the behavior of many ferroelectric materials under applied fields can be qualitatively interpreted by considering relatively simple interactions between DWs and charged point defects. This is particularly true for high-field P-E (switching) hysteresis (see, e.g., reference [18]) and, to some degree, for subswitching responses (see, e.g., reference [4]). Precaution is, however, necessary because there are various parameters to consider, such as type, concentration, location and mobility of defects along with their binding state, as well as the presence of other pinning centers, including grain boundaries [12] or oxygen octahedra tilts [15]. Nevertheless, a simple exercise linking possible DW-defect interactions with the macroscopic response, which will be explained next, is very useful to assess the behavior of a given ferroelectric material.

Early studies identified three microscopic scenarios to describe the strong pinning and DW stabilization effects observed macroscopically by pinched and/or biased high-field P-E hysteresis loops of acceptor-doped PZT and $BaTiO_3$ [8–11, 17]. The three scenarios are illustrated in **Figure 1a** and assume different locations of the pinning centers: (i) inside the domains (called "volume" or, sometimes, "bulk" effect), (ii) in DW regions ("domain wall" effect) and (iii) in or close grain boundary regions ("grain boundary" effect). The common feature of all these scenarios is that defects are arranged in a somewhat ordered manner and, by that, they stabilize the domain configuration via electric and elastic coupling [11, 18], inhibiting domain wall motion. Ordered defects are a characteristic of the so-called "hard" ferroelectrics, exemplified by acceptor-doped PZT. In contrast, "soft"

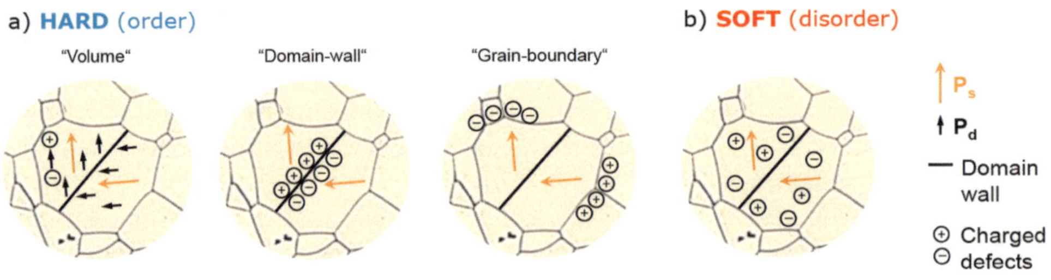

Figure 1.
Simplified microscopic pictures of interactions between charged point defects and domain walls in (a) hard and (b) soft ferroelectrics. P_s and P_d are spontaneous and defect polarization, respectively. Thick black lines denote DWs and circles noted with + and – signs represent individual charged point defects (see also legend on the right-hand side of the schematics). To be further noted is that P_d represents the well-known acceptor–oxygen-vacancy defect complexes (see text for details). The pictures shown in panel (a) were adapted according to the pioneering studies by Jonker [10], Carl and Hardtl [9], Lambeck and Jonker [8] and Robels and Arlt [11]. The scenario in panel (b) assumes disordered defects in soft materials as discussed in, e.g., reference [4].

ferroelectrics, represented by donor-doped PZT, are usually described by assuming disordered defects as shown in **Figure 1b**. Hard-soft transitions, induced in hard-type compositions by disordering the otherwise ordered defects, can be achieved by either electric-field cycling [9, 60] or quenching [30, 61, 62].

The "volume scenario" shown in **Figure 1a** (left schematic), which assumes binding of defects into complexes (see black arrows), is probably the only DW-pinning mechanism that is highly accepted in our research community and is widely used to explain hardening and aging in acceptor-doped PZT, $BaTiO_3$ and similar perovskites (an excellent review on this topic can be found in the paper by Genenko et al. [63]). DW pinning effects mediated by orientation of defect complexes have been put in a theoretical framework more than 30 years ago [17]. More recent studies confirmed the formation of defect complexes and their orientation kinetics both theoretically (by first principles, [64]) and experimentally. In particular, electron paramagnetic resonance (EPR) spectroscopy was found to be a powerful tool due to its high sensitivity, not only to the binding state of certain defects (for details see [65]), but also to the orientation of the defect complexes in the material [65, 66], allowing thus to directly monitor the alignment kinetics of the complexes under the applied field [67]. In most cases, the identified complexes are of the acceptor–oxygen-vacancy type. The literature contains many examples of lead-based and lead-free ferroelectric compositions where such complexes have been identified by EPR; the list includes (but is probably not limited to) $PbTiO_3$, $PbZrO_3$, PZT, $BaTiO_3$, KNN and $(Bi,Na)TiO_3$ (BNT). The dopants in those cases are typical B-site acceptors, such as Fe, Cu and Mn [65, 66, 68–71].

The "DW scenario", which assumes defects accumulated in DW regions (**Figure 1a**, middle schematic), was first predicted by Postnikov more than 50 years ago [72]. Note that the original model does not assume defects bound into complexes. Interestingly, despite being seriously discussed in studies on fatigue mechanisms [73], the DW scenario received strong criticism (to be even labeled as "a well-known speculation") from those that were strictly in favor of the volume scenario [74]. Nevertheless, very recent studies on BFO ceramics using atomic-resolution electron microscopy, which will be discussed later, support the idea of this scenario and indeed suggest that is active in this material [75].

The "grain-boundary scenario" (**Figure 1a**, right schematic), often referred to as the space-charge mechanism, was extensively discussed in a combined theoretical and experimental frame against the volume scenario originally proposed by Arlt et al. [17]. Key hardening characteristics of acceptor (Fe) doped PZT, such as the dopant-concentration dependent aging time, were shown to be well predicted by the space charge model [76, 77]. A perhaps interesting observation is that phenomenological calculations revealed that the electrostatic pinning effects on DWs from the space charges can be two orders of magnitude stronger than the pinning effects provided by the aligned defect complexes (for the same charge-carrier concentration) [76]. Considering that grain boundaries are strained areas and sources of polarization discontinuities where point defects may be attracted due to electrostatic and elastic driving forces, the grain boundary scenario should probably be always considered when interpreting the macroscopic responses of ceramics to external driving fields, particularly if the ceramics are fine-grained.

It is clear that more than one of the three domain-stabilizing effects shown in **Figure 1a** may be active in a given material. Actually, the reason for considering only a single scenario is in the human nature seeking for the simplest explanation. A number of arguments supports the idea of multiple pinning mechanisms. I give two examples. In hard PZT, it was predicted by first principles that the acceptor–oxygen-vacancy defect complexes should energetically prefer to be situated closer to DWs (as shown schematically with black arrows in **Figure 1a**, left schematic) [78]. This means that the pinning mechanism may consist of a combination of volume and DW effects. In another study based on conductive atomic-force microscopy (c-AFM) analyses, it has been shown that in BFO ceramics both DWs and grain boundaries exhibit enhanced electrical conductivity with respect to the bulk conductivity measured in the grain interiors [79]. This suggests the presence of mobile charges at both locations, which may act as pinning centers. In this same material, defect complexes based on oxygen vacancies have been also identified by EPR [80]. The overall data, therefore, point to an extremely complex situation where all three pinning scenarios shown in **Figure 1a** may be active. The challenge is to find whether and which mechanism is dominant. At present, conductive DWs appear to be the key features affecting the piezoelectric response of BFO, which will be discussed in detail in the next section.

In contrast to ordered defects in hard ferroelectrics, disordered defects in soft counterparts, as shown schematically in **Figure 1b**, are more difficult to be directly identified. Nevertheless, in the case of PZT, for example, the presence of disorder in donor-doped samples (in a pragmatic sense) is supported by the fact that the measured macroscopic subswitching nonlinearity and hysteresis are well consistent with the major predictions from the Rayleigh model [27]. Another argument is that EPR data on Gd-doped $PbTiO_3$ indicate no binding of the donor (Gd) dopant with the expected compensating Pb vacancies [81], supporting the results of first-principles studies on the same perovskite [78]. In addition, none of the two defects (donor dopants and Pb vacancies) are mobile below the Curie temperature of the material, where the defects can in principle be ordered because they are provided by electric and elastic driving forces originating from the spontaneous polarization and strain, respectively (see, e.g., the graphic explanations by Ren [18]). If the two defects are not mobile nor they show tendency of binding, then a "frozen-in" defect disorder state in donor-doped PT or PZT can be envisioned. It is not unusual, however, that the dopants segregate at the grain boundaries (a nice recent example is shown for Ti-doped BFO in Ref. [82]). The issue of defect segregation at or close to grain boundary regions should be more seriously considered, as recently pointed out by Slouka et al. [83].

Other discussions related to the nature of defects in soft PZT point to the likely possibility that the donor dopant reduces the concentration of oxygen vacancies in PZT (due to charge compensation), leading to a progressive transition from a state characterized by ordered defects (undoped PZT) to a state governed by disordered defects, as the donor dopant is added to PZT [29]. In this sense, experimental data even indicate that the dominant pinning centers affecting DW displacements and, consequently, nonlinearity, in both hard and soft PZT, could be oxygen vacancies. The lesson that can be learned from all these data is that the defect arrangements shown in **Figure 1** should not be treated individually; as a matter of fact, all of them may be present, to some degree, in a given ferroelectric. Finally, it has to be emphasized that the situation in soft materials is more complex, and the true origins of softening are still not clear (if the reader is interested in these issues, it is recommended to consult the work of Dragan Damjanovic [29, 78]).

3. Piezoelectric nonlinearity and hysteretic response of BFO ceramics: low-frequency nonlinearity and negative piezoelectric phase angle

One of the distinct characteristics of the piezoelectric response of BFO is the strong dependence of the nonlinearity on the frequency of the driving field. This is illustrated in **Figure 2**, which compares the converse piezoelectric response of BFO with that of hard and soft PZT. The PZT compositions correspond to the rhombohedral symmetry and were selected exactly to enable a comparison with the isostructural rhombohedral BFO, although symmetry is probably the only common aspect of these two perovskites. In the first place, it should be mentioned that the increasing tendency of d_{33} with the electric-field amplitude E_0 observed in **Figure 2a–c** for the two PZTs (at all measured frequencies) and BFO (mostly at 1 and 0.2 Hz) is accompanied by an increase in the piezoelectric tanδ (not shown here). This suggests irreversibility in the response as implied by the intimate nonlinearity-hysteresis relationship predicted by RL (see Eqs. (1) and (2)). The nonlinear behavior in these samples can be thus related to irreversible domain wall contribution. This is supported by in situ XRD analysis on BFO [84] and similar PZT samples to those shown here [26].

As expected and described in the previous section, due to stronger DW pinning effects, hard PZT shows a lower absolute d_{33} value, which is less dependent on the field magnitude (**Figure 2a**) than that of soft PZT (**Figure 2b**). Also, the d_{33} field

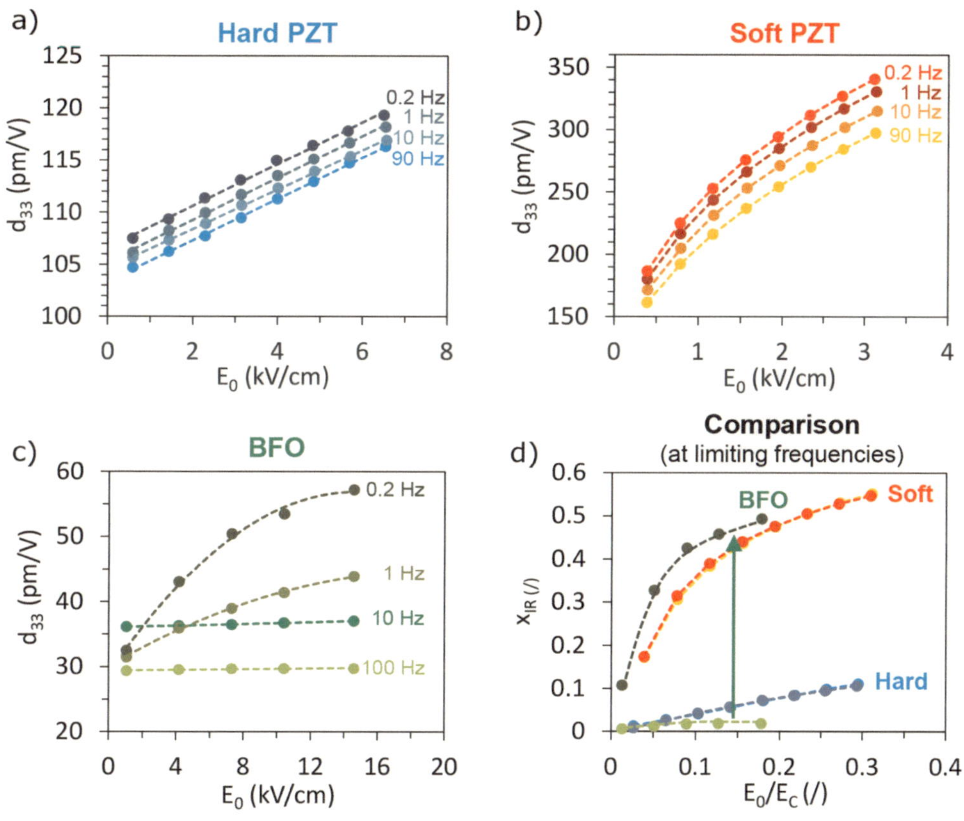

Figure 2.
Converse piezoelectric d_{33} coefficient as a function of electric-field amplitude E_0 measured at different frequencies for (a) 1% Fe doped $Pb(Zr_{0.6}Ti_{0.4})O_3$ (hard PZT), (b) 1% Nb doped $Pb(Zr_{0.6}Ti_{0.4})O_3$ (soft PZT) and (c) BFO ceramics. (d) Comparison of the fraction of the piezoelectric d_{33} coefficient related to the irreversible contribution (X_{IR}; see Eq. (3)) among the three samples. The response is in this case shown for two limited frequencies (90 or 100 Hz and 0.2 Hz) and as a function of relative electric field, i.e., E_0/E_C (E_C is the coercive field determined by standard P-E hysteresis measurements, which are not shown here). For clarity, color coding of the curves is the same in all panels. The green arrow in panel (d) illustrates the strong effect of the driving electric-field frequency on the nonlinearity in BFO, which is absent in hard and soft PZT. The data correspond to fine-grained BFO sample (average grain size 1.8 µm) as reported in Ref. [79].

dependency in soft PZT is sublinear with respect to the rather linear dependency in hard PZT. This behavior is retained at all frequencies used in the measurements. The sublinear trend in soft PZT has been discussed to some degree in the introduction of this chapter in relation to non-uniform pinning potential.

The next information can be obtained by quantifying the data. The slope of the d_{33}-vs-E_0 curves is for hard PZT $\sim 0.2 \cdot 10^{-16}$ m^2 V^{-2} and is practically independent on the frequency. By contrast, in soft PZT, the slope (in this case calculated for 1 kV cm^{-1}, which corresponds to 10% of coercive field E_C, i.e., $E/E_C = 0.1$) is an order of magnitude higher, i.e., $8.1 \cdot 10^{-16}$ m^2 V^{-2} at 90 Hz, and further increases to $9.9 \cdot 10^{-16}$ m^2 V^{-2} at 0.2 Hz. The frequency dependence of the irreversible coefficient of soft PZT has been reported earlier and correlated with interface pinning in a disordered medium [85].

As easily assessed from **Figure 2c**, the nonlinearity and the associated irreversible contribution to piezoelectricity in BFO is completely different than in PZT. Here, the nonlinear behavior is strongly dependent on frequency: the d_{33}-vs-E_0 slope is nearly zero at 100 Hz, precisely $0.003 \cdot 10^{-16}$ m^2 V^{-2}, and increases to $0.27 \cdot 10^{-16}$ m^2 V^{-2} when the frequency is reduced to 0.2 Hz (as in the previous case on PZT, the slope was calculated for the relative field $E/E_C = 0.1$). Such a dramatic two-orders-of-magnitude increase in the nonlinearity coefficient with reduced frequency is obviously absent in the two PZT variants.

A direct comparison between the nonlinearity of BFO and PZT is shown in **Figure 2d**. As a quantitative measure, we use the fraction of the total piezoelectric coefficient that is due to irreversible DW displacements (X_{IR}). This parameter is essentially represented by the fraction of the field-dependent coefficient and can be derived from Eq. (1) as:

$$X_{IR}(E_0) = \frac{d_{33}(E_0) - d_{33}(E_0 = 0)}{d(E_0)} \qquad (3)$$

where $d_{33}(E_0)$ and $d_{33}(E_0 = 0)$ is the coefficient at a given field amplitude E_0 and at zero field amplitude, respectively. As anticipated, the irreversible contribution in hard and soft PZT is practically independent on the frequency (**Figure 2d**, blue and red data). X_{IR} reaches a maximum of 10% and 55% in hard and soft PZT, respectively. Interestingly, the irreversible contribution in BFO at 100 Hz is smaller than that of hard PZT and becomes higher at 0.2 Hz than that of soft PZT (for the same relative field E/E_C; see green data in **Figure 2d**). Note that the large contribution from the displacements of non-180° domain walls at sub-Hz driving frequencies in BFO was recently confirmed by in situ XRD stroboscopic analysis [84]. One could imagine BFO as very hard material at high frequencies and very soft at low frequencies. The data, therefore, present a distinct hard-to-soft transition in BFO induced by the driving frequency. As will be explained throughout the rest of the chapter, this transition originates from the presence and dynamics of conductive DWs.

Another distinct feature of the piezoelectric response of BFO is the negative piezoelectric phase angle (here and in subsequent discussion, the phase is represented as the tangent of the piezoelectric phase angle, $\tan\delta_p$). This rather unusual response is particularly strong in coarse-grained BFO and is presented in **Figure 3**. From these data it is clear that the piezoelectric response of BFO is rather complex and show strong frequency dependence (**Figure 3a**). When inspected from high to low frequencies (right to left), the response can be described as a sequence of increasing and decreasing d_{33} in the frequency range 1–200 Hz and 0.1–1 Hz, respectively, giving

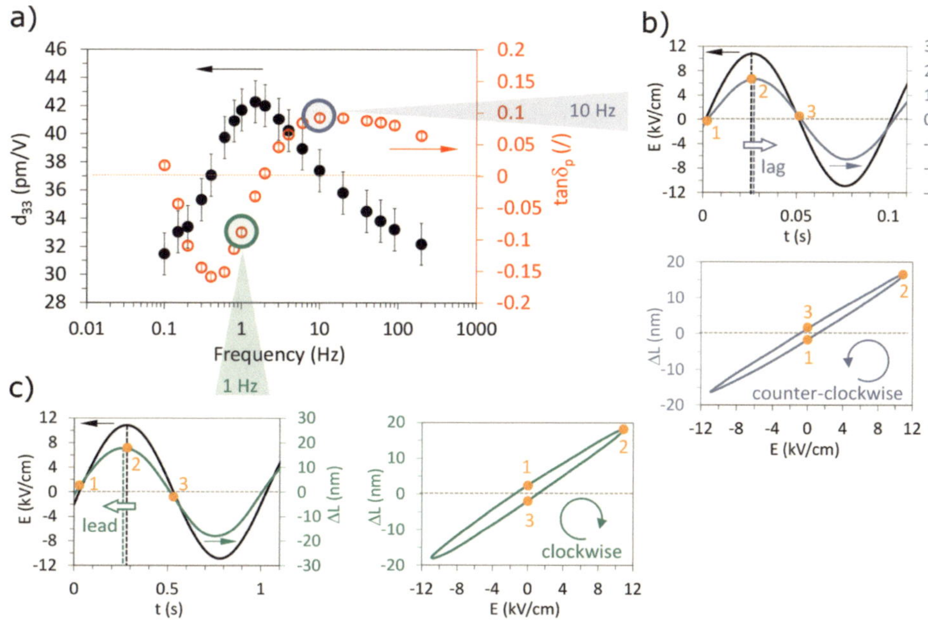

Figure 3.
(a) Piezoelectric coefficient (d_{33}) and tangent of the piezoelectric phase angle ($tan\delta_p$) as a function of driving field frequency for coarse-grained BFO. The data were obtained at 11.9 kV/cm of driving field amplitude. Error bars represent measurement error. These errors alongside the details of the annealing procedure of the coarse-grained BFO sample (average grain size 16 μm), its microstructure, domain structure and local electrical conductivity are reported in Ref. [79]. The two insets show the time-domain driving field E signal (black sinusoidal curves) with overlaid mechanical displacement ΔL signals (blue or green sinusoidal curves) together with the corresponding ΔL-E piezoelectric hysteresis loops for the case of (b) positive (10 Hz; blue data) and (c) negative (1 Hz; green data) piezoelectric phase angle. The bigger arrows in panels (b) and (c) indicate the lagging and leading ΔL signal with respect to the driving field signal for the case of positive and negative phase angle, respectively. The corresponding counter-clockwise and clock-wise rotational sense of the hysteresis, which is set by the lagging and leading output signals, respectively, is also noted on the respective loops. Numbered points on the time-domain plots and hysteresis are drawn to help identifying the rotational sense of the hysteresis.

rise to a d_{33} maximum at 1.5 Hz (**Figure 3a**, black points). These two d_{33}-vs-frequency behaviors are accompanied by a broad +$\tan\delta_p$ maximum at ~10 Hz and a rather narrow –$\tan\delta_p$ minimum at 0.4 Hz, respectively (**Figure 3a**, red points). The overall behavior can be thus simplified to a sequence of retardation process (d_{33} increasing with decreasing frequency accompanied by +$\tan\delta_p$ peak) and relaxation process (d_{33} decreasing with decreasing frequency accompanied by –$\tan\delta_p$ peak) [4, 46]. As it will be shown in the next section, this is consistent with the distinct features of M-W piezoelectric effect, which can be easily predicted by simple analytical modeling.

As it has been done in the case described here, the sign of the piezoelectric phase is very often determined by measurements performed using lock-in technique, which is capable of extracting the amplitude and phase of individual harmonic responses to external sinusoidal excitations [29]. For less experienced, this may be sometimes non-trivial as several instruments may reverse (by 180°) or somewhat affect the phase of the output signal. While the sign of $\tan\delta_p$ can be checked by, e.g., measurements of a sample with known response, such as, e.g., a donor-doped soft PZT where $\tan\delta_p$ should be positive (and, ideally, related to Eqs. (1) and (2)), it is useful to acquire the total signal containing all information by conventional oscilloscope imaging. Two examples of such acquired signals displayed either in time domain or as piezoelectric hysteresis loop (mechanical displacement ΔL versus electric field E plots) are shown in **Figure 3b** and **c**. The case in **Figure 3b** is shown for 10 Hz where a positive $\tan\delta_p$ was measured using the lock-in method. As expected, the output displacement ΔL signal lags behind the input driving field E signal (see arrow in **Figure 3b**), corresponding to counter-clockwise rotational sense of the piezoelectric hysteresis. In terms of hysteresis, this is a common situation observed during, e.g., conventional measurements of ferroelectric P-E hysteresis loop at switching fields (to give a popular example). The case that is less common is when $\tan\delta_p$ is negative as shown in **Figure 3c**. Here, instead of *lagging*, the output ΔL signal *leads* the input E signal, effectively resulting in a clock-wise rotation of the piezoelectric hysteresis.

Very often (and not to be blamed) the negative piezoelectric phase is misinterpreted because it gives a wrong impression that it violates the basic law of energy conservation. This is also the reason why is so often interpreted as a measurement artifact. An example is the clockwise hysteresis measured in the response of ferroelectric field-effect transistor thin-film structures, which arise due to charge injection during measurements [86, 87].

It is not within the scope of this contribution to provide a deep and detailed physical analysis of the negative piezoelectric phase, neither is such analysis within the main expertise of the author of this chapter. The reader is strongly advised to follow the discussion provided in the chapter on hysteresis by Damjanovic [4]. Nevertheless, one can envision a very simple reasoning based on the classical power dissipation principles in dielectrics (found in general textbooks, such as [88]) where the power loss is defined by the positive dielectric $\tan\delta$ (considering that all other parameters in the equation for the dissipated power density, i.e., electric-field amplitude, frequency and real part of the dielectric permittivity, are positive by definition). Similarly, the area of the charge density (D) – electric field (E) hysteresis represents the energy loss of a dielectric during an electric field cycle per unit volume of the material (units of D-E hysteresis is J m^{-3}) [89]. In this analogy, one could understand a negative piezoelectric phase angle corresponding to energy gain that is represented by the piezoelectric hysteresis area. The simplest and perhaps strongest argument against this claim is that, unlike the dielectric D-E hysteresis area, the piezoelectric hysteresis area (either converse x-E or direct D-Π) does not have units of energy density. Therefore, the hysteresis shown in **Figure 3c**, where the phase is negative, does not *directly* represent an energy gain because the hysteresis area does not reflect energy density. This problem has been extensively elaborated by Holland [90]. The rigorous mathematical treatment therein shows that the necessary restriction of positive total power loss, which is represented by the sum of dielectric, elastic and mix piezoelectric components (proportional to respective imaginary coefficients), results into dielectric and elastic loss terms being always positive and bigger than the piezoelectric term. The latter was shown to not be restricted in its sign. Therefore, taking the longitudinal piezoelectric response of poled ferroelectric ceramics as an example, the imaginary piezoelectric coefficient d_{33}'' is permitted to be either positive or negative; for a positive real longitudinal coefficient d_{33}', as is the case of ferroelectric ceramics, the piezoelectric $\tan\delta_p$ (defined as $\tan\delta_p = \frac{d''_{33}}{d'_{33}}$) can thus be either positive or negative. While not restricted in its sign, the piezoelectric term in the total power dissipation function is, however, restricted in magnitude. This essentially means that a negative piezoelectric phase angle measured in poled ceramics indicates a partial reduction of the total power loss. To avoid confusion, it should be noticed that in the power dissipation equation, reported in the paper by

Holland, the term related to the piezoelectric coupling is proportional to E·Π·d", which must obviously possess units of energy density (subscripts are omitted for simplicity). In contrast, the piezoelectric hysteresis area is proportional to either the product of x·E (converse effect) or D·Π (direct effect), none of which possess energy density units. A very nice experimental example of the reduction of the total power dissipation due to piezoelectric coupling was shown for Sm-doped $PbTiO_3$ [4].

Coming back to our case on BFO, it must be emphasized at this point that the negative converse piezoelectric phase angle has also been confirmed in the direct piezoelectric response of BFO [91]. In addition, a negative phase has also been detected in the lattice microstrain response of BFO ceramics to external electric field,

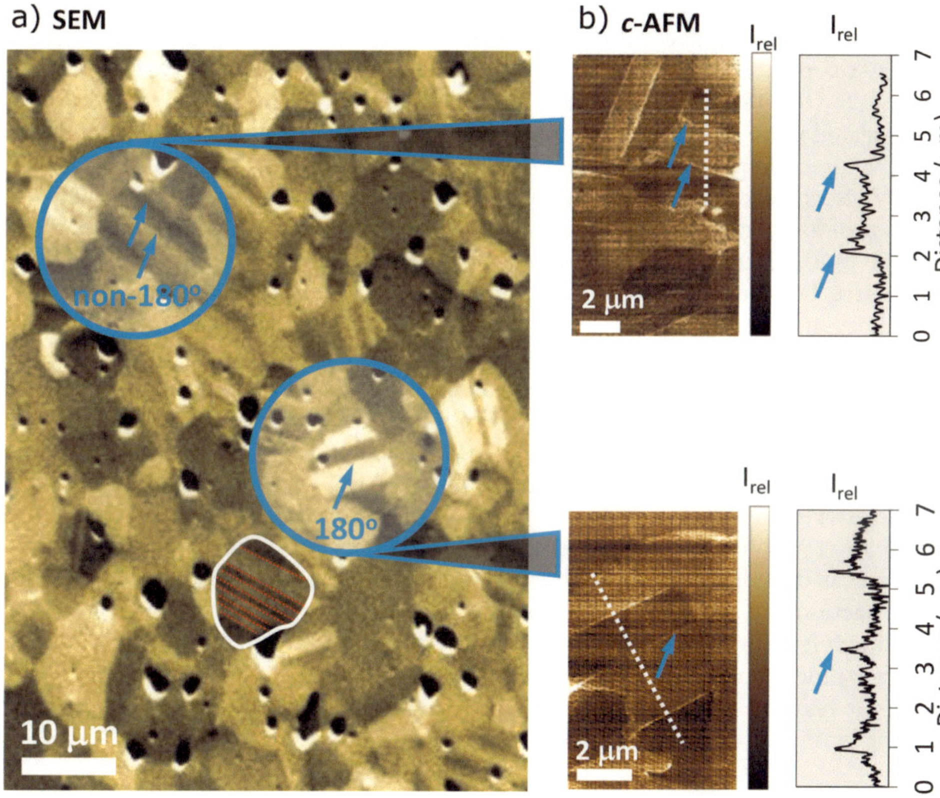

Figure 4.
(a) Scanning electron microscopy (SEM) image of the surface of BFO ceramics with visible grains and domains. At the lower part of the image, a grain was highlighted for clarity with the thick white line indicating the grain boundary and the thin red dashed lines indicating DWs. Note that the domains, which are separated by these DWs, are clearly seen as alternating dark/bright bands in the SEM image. The other two regions highlighted with blue circles were further analyzed by means of EBSD analysis. In each of the two regions, the type of DW (either 180° or non-180°) was determined based on the orientation of [111] polar axis of the rhombohedral lattice (with respect to the surface sample plane) in domains adjacent to the analyzed DW (for details, see text and reference [79]). (b) c-AFM maps of the same regions which are in panel (a) highlighted with blue circles. I_{rel} on the scale bar signifies relative current, displayed with respect to the average background current signal. The bright lines observed at DWs (see blue arrows on the maps), which correspond to an increased local electric current signal, confirm the enhanced conductivity at both 180° (bottom c-AFM image) and non-180° DWs (top c-AFM image). Adjacent to the c-AFM maps are the electric current profiles measured by scanning the AFM tip on the surface along the white dashed lines noted on respective c-AFM maps. The peaks noted by blue arrows clearly emerge from the electric-current background, supporting the idea of electrically conducting 180° and non-180° DWs in BFO ceramics. Parts of the figure are reprinted from reference [79] with the permission of John Wiley and Sons.

which was characterized by in situ XRD stroboscopic analysis [84]. This is perhaps the most important evidence of a negative piezoelectric phase angle measured in any piezoelectric so far because, prior to that work, such response (to the best of the author's knowledge) was demonstrated only on the level of macroscopic measurements.

As anticipated in the initial discussion of the results shown in **Figure 3**, the negative piezoelectric phase angle is a strong indication of M-W piezoelectric process that is very common in, e.g., piezoelectric composites [46, 47]. In analogy to the dielectric M-W relaxation, characterized by giant apparent dielectric permittivity [50], the modeling of Turik et al. showed the same effects in the piezoelectric M-W analogue [45]. As explained earlier, the M-W relaxation has origin in the internal electric field redistribution in a medium, where local regions exhibit different electrical conductivities; in the frame of modeling, such regions are often implemented in the form of M-W bilayer units, where the two layers are described by different conductivities [48, 51]. This is the reason behind a large piezoelectric M-W effect in some Aurivillius phases, such as $Bi_4Ti_3O_{12}$ [48]. These ceramics, in fact, tend to show anisotropic microstructure with elongated plate-like grains that are characterized by different electrical conductivities in the direction parallel or perpendicular to the plane of the plates. The piezoelectric M-W effect in heterogeneous $Bi_4Ti_3O_{12}$ ceramics is thus easy to support using arguments of anisotropic conductivity.

Unlike in $Bi_4Ti_3O_{12}$, the M-W features observed in the piezoelectric response of BFO (**Figure 3**) are more difficult to interpret because significant anisotropy in the conductivity in a homogeneous perovskite oxide, such as BFO, is not expected, at least not to a level as in layered Aurivillius-type structures. Also, BFO is characterized by a microstructure typically consisting of equiaxial grains (as it is illustrated in **Figure 4** in the next section). It should be recalled, however, that the piezoelectric M-W effect in BFO is particular in that it shows, in addition to the negative piezoelectric phase angle (**Figure 3**), a very strong DW contribution observed only at low (sub-Hz) driving frequencies (**Figure 2c**). The overall data thus suggest a piezoelectric M-W effect provoked by DWs. The idea becomes reasonable when DWs are electrically conducting as the conductivity is what triggers the M-W effect. As it will be shown in the next section, it is exactly the conducting DWs that likely cause large anisotropy in the conductivity from grain to grain or across cluster of grains. Obviously, in BFO the situation is more complex than in other known piezoelectric M-W cases, simply because the features that are triggering the M-W effects, i.e., the DWs, can also move inside the grains under applied fields.

4. Analytical modeling of nonlinear Maxwell-Wagner piezoelectric relaxation arising from electrically conducting DWs

While, pragmatically speaking, the idea of conducting DWs giving rise to macroscopic piezoelectric M-W effects does fit into a reasonable explanation, the details of the mechanism are rather complicated. The reason is the complex situation in a polycrystalline ceramic matrix containing randomly oriented grains, each characterized by a domain structure with different DWs and corresponding DW planes oriented in various directions with respect to the reference external field axis. The DW planes are assumed to provide conductive paths through individual grains or in local regions inside the grains, modifying the internal electric fields. If further grain-to-grain elastic interactions are considered, which have a significant effect in polycrystalline piezoceramics [26], modeling may become extremely difficult. In the first approximation, therefore, the model can be simplified and reduced to a level that allows the major macroscopic parameters to be predicted and compared with experimental data. As demonstrated for the case of direct piezoelectric response of Aurivillius phases [48], analytical bilayer modeling is sometimes sufficient. A similar case study on BFO, but additionally elaborated to incorporate nonlinear effects, will be presented in this section.

In the first place, it is important to work out a physical picture onto which to construct the model. Following the original discovery of conducting DWs in epitaxial BFO thin films [92], conduction at DWs in BFO polycrystalline samples have been determined using the same c-AFM measurements, which make it possible to probe the local conductivity of the sample. In BFO films, the epitaxial growth determines the crystallographic orientation of the film with respect to the substrate plane, so that the determination of the type of DWs in rhombohedral (R) BFO (71°, 109° or 180°) is facilitated. In ceramics, this is more difficult as the orientation of the grains is random and it is thus not known when imaging the surface with microscopy techniques. A simple way to accomplish the task is to determine the orientation of the polar [111] axis of the BFO R symmetry in adjacent domains in a selected grain using, e.g., electron back-scattered diffraction (EBSD) analysis. By indexing the Kikuchi patterns obtained in individual domains, as those highlighted in blue circles in **Figure 4a**, with the available R BFO space group, the angle between the [111] vectors in adjacent domains can be determined. Once these angles are known and without going into detailed geometrical analyses, which may be non-trivial (see the example in the supplementary material 1 of the paper [93]), a separation between 180° and non-180° DW-type is possible. In the former case, the angle should be zero ([111] vectors in adjacent domains are parallel), while in the second is non-zero. Note that further analysis is complicated by the fact that EBDS cannot determine the orientation of the ferroelectric spontaneous polarization, but only the ferroelastic distortion.

EBSD analysis was used to determine the 180° and non-180° DWs in the example shown in **Figure 4a** (see also reference [79] for further details). In the next step, these same regions were analyzed by c-AFM (**Figure 4b**) to finally confirm that both 180° and non-180° DWs in R BFO ceramics exhibit enhanced electrical conductivity with respect to that measured in the domains (see also the electric-current profiles in **Figure 4b** measured across DWs as indicated with the dashed white lines in respective c-AFM maps). The results are fully consistent with the atomic-scale microscopy analysis, which confirmed the presence of p-type charge carriers, identified as Fe^{4+} states, concentrated inside all three DW variants (71°, 109° and 180°) of BFO [19], resulting in the domains mostly likely depleted from the charges. The p-type conductive nature of the DWs is supported by annealing studies in controlled atmospheres, reported in the same paper, while the dynamic pinning effects of the p-type carriers is reported in Ref. [75].

The analysis of the effect of conductive DWs on the piezoelectric response can be reduced, in the first approximation, to a two-grain problem as shown in **Figure 5a** (left schematic). Since ceramics are composed of randomly oriented grains containing different type of DWs with various orientations, it is legitimate to analyze a hypothetical case of two grains oriented along [100] and [111] directions with respect to the externally applied electric field vector. The choice of these orientations will become evident in the subsequent discussion. If the two grains contain only 71° DWs percolating the grains, then the situation should closely resemble the left schematic shown in **Figure 5a** (detailed geometrical analysis of the DW angles in the two grains is reported elsewhere [84]). The key element of the model is the different orientation

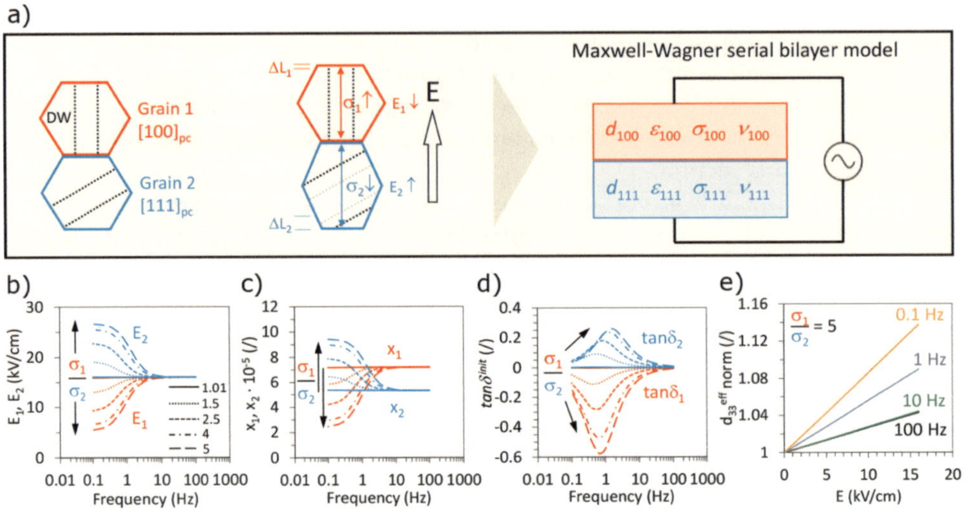

Figure 5.
(a) Schematic of the two-grain model (left) representing a serial bilayer M-W unit (right). The two grains are assumed $[100]_{pc}$ and $[111]_{pc}$ oriented with respect to field axis E and are denoted as grain 1 (red) and grain 2 (blue), respectively. Notation "pc" indicates "pseudocubic". The dashed lines inside the grains represent DWs. The DW orientation is assumed such that the DW planes in grain 1 are parallel to E, while those in grain 2 are not, forming an angle of ~35° with respect to E axis. These angles should correspond to those of 71° DWs present in the two grains (for detail geometrical analysis, see reference [84]). Due to the different orientation of the electrically conductive DW in the two grains, the conductivity measured along E axis will be higher in grain 1 than in grain 2 (see σ_1 and σ_2 notation with upward and downward arrow in grain 1 and grain 2, respectively). This will result in the drop of the electric field in grain 1 (noted by down-arrowed E_1) and rise of the electric field in grain 2 (up-arrowed E_2). The same difference in the orientation of the DW with respect to E will result in a lattice strain with no contribution from DW motion in grain 1 (see ΔL_1), while displacements of DWs is expected in grain2, in addition to lattice strain (see larger ΔL_2). To incorporate DW contributions, the calculations were performed assuming RL relations in grain 2 as described in detail in Ref. [94]. The two grains may be viewed as a M-W bilayer unit where each layer is assumed as a leaky piezoelectric element, characterized by its own piezoelectric coefficient d_i, dielectric permittivity ε_i, specific DC electrical conductivity σ_i and volume fraction ν_i (i = 100 or 111 indicating the two grain orientations). Panels (b–d) show driving-frequency dependent (b) internal electric field distribution (E_1, E_2), (c) piezoelectric microstrains (x_1, x_2) and (d) tangent of the piezoelectric phase angle ($\tan\delta_1$, $\tan\delta_2$) of the two grains (layers) as calculated from the model. The different curves in panels (b,c,d) correspond to different anisotropic parameter, i.e., ratio of the conductivities $\frac{\sigma_1}{\sigma_2}$, which were set to 1.01, 1.5, 2.5, 4 and 5 (the increasing tendency of $\frac{\sigma_1}{\sigma_2}$ is shown by arrows on individual plots and the corresponding curves are drawn with different lines, each corresponding to specific $\frac{\sigma_1}{\sigma_2}$ ratio as noted in panel b). The data in panels (b,c) were calculated for 16 kV/cm of externally applied sinusoidal field amplitude, while those shown in panel (d) were calculated for zero-field amplitude (linear piezoelectric $\tan\delta^{init}$). (e) Effective (total) piezoelectric coefficient of the M-W bilayer unit normalized to the value at zero field ($d_{33}{}^{eff}$ norm) as a function of external (nominal) electric field amplitude calculated for different driving field frequencies (the anisotropic parameter $\frac{\sigma_1}{\sigma_2}$ was fixed to a value of 5). Parts of the figure are reprinted from Ref. [94] with the permission of AIP Publishing.

of conductive DWs in the two grains with respect to the external field axis, i.e., in the top grain 1 (red) the DWs are parallel to E, while in the bottom grain 2 (blue) they form a different angle. The net result is that the conductivity, measured along E, of grain 1 should be higher than that of grain 2 because in the former the charges may migrate along the vertical conductive DWs (see notations σ_1 and σ_2 in **Figure 5a** indicating the conductivity measured vertically in grain 1 and 2, respectively). Using the same reasoning, the conductivity of grain 2 should be higher when measured along a direction away from the E axis (not parallel). It is this anisotropy in the electric conductivity that it is assumed here to lead to M-W-like internal field redistribution (see down-arrowed E_1 and up-arrowed E_2 notations in **Figure 5a**, denoting the internal fields). Note also that the bottom grain 2 is oriented in a way to give rise to a stronger DW contribution than the upper grain 1; in the ideal case, due to orientational constraints, grain 1 should exhibit only lattice strain as a response to the external E (in **Figure 5a** the microstrains arising from the two grains due to field application are noted as ΔL_1 and ΔL_2). Further details regarding the model and mathematical derivations can be found in the supplementary Section 5 of the paper by Makarovic et al. [94].

The assembly of the two grains represents a basic M-W bilayer unit. As in classical dielectric M-W modeling [50], the picture can be rationalized in terms of an equivalent circuit consisting of two leaky capacitors connected in series, where each leaky capacitor may be represented by an ideal capacitor and an ideal resistor connected in parallel (**Figure 5a**, right-hand schematic). Obviously, the piezoelectric effect is added to the two layers (grains). Due to the expected DW contribution in the bottom grain 2, the RL relations (Eqs. (1) and (2)) were used to calculate the piezoelectric response of this grain. In contrast, grain 1 is assumed to respond via intrinsic lattice piezoelectric effect, so its response can be modeled by the linear constitutive piezoelectric equation. For a set of dielectric, piezoelectric and conductivity parameters, which are within the margins reasonable for BFO (see reference [94]), the results of the modeling are shown in **Figure 5b–e**.

Driven by the anisotropy in the electrical conductivity, defined as the conductivity ratio of the two grains ($\frac{\sigma_1}{\sigma_2}$), the internal fields in the grains (E_1, E_2) will be redistributed as a function of driving field frequency (**Figure 5b**). In accordance to the conductive behavior related to the different DW orientation in the two grains, the sinusoidal electric field applied externally to the bilayer serial structure will be redistributed at low driving frequencies (<10 Hz) such that the electric field in the top grain 1 (red color coding) will be reduced with respect to the nominal (external) field, while that of the bottom grain 2 (blue) will be increased instead (**Figure 5b**). In other words, due to leakage in grain 1 caused by the vertical orientation of conductive DWs, the field inside this grain will drop and will be thus transferred from the leaky grain 1 to the less-leaky grain 2 (see **Figure 5a**). This will happen at low driving frequencies as such driving conditions provide to the charges sufficient time to migrate along conductive DWs. Being proportional to the internal field, the microstrains of the two grains (**Figure 5c**) will show the same frequency behavior as that of respective internal fields, leading to either retardation and a peak in the positive piezoelectric phase angle (see $\tan\delta_2$ in **Figure 5d**), or relaxation and a peak in the negative piezoelectric phase angle (see $\tan\delta_1$ in **Figure 5d**). This M-W mechanism will ultimately lead to nonlinear effects: the increased internal electric field in grain 2 (E_2 in **Figure 5b**) will boost the DW motion in this grain, resulting in effective nonlinearity enhanced at low driving frequencies as shown in **Figure 5e** (compare also with experimental data, **Figure 2c**). The piezoelectric M-W effect has thus a nonlinear character or, in other words, the piezoelectric nonlinearity (i.e., DW motion) is boosted due to the electrical conductivity.

At this stage, I have to point out that the presented model used to understand the experimental data on BFO (as those shown in **Figures 2c** and **3**) is largely simplified: (i) it considers only two isolated and unconstrained grains connected in series, neglecting the true elastic and electric boundary conditions of the analyzed grains set by the presence of other surrounding grains, (ii) it does not consider elastic coupling of the two isolated grains and transverse piezoelectric effects (as was done in, e.g., reference [51]), (iii) it assumes very simple domain structure in the two grains and only one type of DW (i.e., 71°), and (iv) the conductivity of the grains set by the conductive paths along DWs is assumed to be fixed, although a dynamic conductivity is not unreasonable considering that the conductive DWs may switch locally (through a RL-like irreversible jump, for example), resulting in a modified conductive path through the switched DW. Despite all these limitations, however, it is surprising to realize that the simple two-grain model predicts all the key experimental observations. First, the model predicts both retardation and relaxation processes (**Figure 5c** and **d**), which are likely convoluted in the experimental response (as discussed for **Figure 3a**). Second, the model shows that the lattice strain response should exhibit a negative phase with respect to sinusoidal external field, thus leading the field signal (**Figure 5d**). Not only is this consistent with macroscopically measured piezoelectric strain leading the field signal (**Figure 3c**), but this is also exactly the behavior of the lattice strain deconvoluted from the total converse piezoelectric response of BFO ceramics using synchrotron XRD measurements [84]. Third, the important outcome of the model related to the low-frequency nonlinearity is a natural consequence of introducing RL relations, lining up with the macroscopic experimental data (**Figure 2c**) and the low-frequency DW contribution determined by synchrotron XRD measurements [84].

The low-frequency nonlinearity has recently been shown to be a response parameter that can be controlled by designing the fraction of conductive DWs in BFO ceramics via doping [94]. The idea was triggered by the outcomes of the model itself, further reinforcing the value and importance of simple modeling. It is probably needless to say that controlling nonlinearity and hysteresis is very important for the development of high-temperature piezoceramics based on BFO. This is indeed supported by a study on BFO showing that the temperature dependent piezoelectric response of these ceramics is strictly controlled by the same M-W processes described in this chapter [95]. Importantly, it was shown that the strong temperature dependence of the piezoelectric nonlinearity and hysteresis, which is of a direct relevance for the device operation, has origin in the thermally activated nature of the local electrical conductivity in BFO ceramics.

Simple analytical modeling was found to be a promising first step toward understanding the complex piezoelectric behavior of BFO arising from conductive DWs. It could be interesting, however, to model a more complex situation to account for the many different parameters that are necessarily neglected in the simple analytical approach. As shown in **Figure 5b–d**, the effect of the anisotropic parameter is crucial as it determines the strength of the M-W effect. For more advanced modeling, this parameter could be viewed as varying from one M-W unit to another in a complex ceramic matrix composed of interacting grains forming units with different time constants (τ is proportional to the ratio of the weighted sum of permittivity and conductivity of the individual layers in the bilayer unit [48, 84]). This could eventually account for the different regions inside the ceramics exhibiting different levels of electric-field redistribution (as shown by the different curves in **Figure 5b**). 3-dimensional (3D) finite element modeling based on a phenomenological approach has been recently demonstrated to be a powerful tool in predicting local electric fields in 3D ceramic matrices with defined porosity [96]. For solving complex problems, such as those encountered in elastically and electrically coupled grains in ferroelectric ceramics, multiscale modeling approaches show great promises [97–99].

Another interesting point that could be considered in the future is to push the relaxation to higher frequencies (equal to decreasing the τ value). This would make it possible to use the large response in a higher frequency range that is more relevant for piezoelectric applications (in the example shown in **Figure 5**, τ = 0.13 s, corresponding to a relaxation frequency of f_{relax} = 1.2 Hz). As discussed in the preceding paragraph, the anisotropy in the conductivity is also an important factor in tailoring the usable frequency range. It is thus tempting to consider designing a matrix with charged DWs, which can possess metallic-like conductivity, as recently demonstrated for $BaTiO_3$ [100]. In this case, the DW conductivity may exceed the bulk conductivity for impressive 8–10 orders of magnitude, perhaps providing an opportunity in designing piezoelectric properties.

5. Summary and conclusions

In this chapter, I have reviewed and discussed a case study on BFO ceramics explaining how the presence and dynamics of DWs showing enhanced electrical conductivity with respect to bulk conductivity can have a crucial effect on the macroscopic piezoelectric response of BFO. The mechanism goes far beyond the expected DW-defects pinning interactions, described in Section 2 of this chapter, and reflect itself in the nonlinear and hysteretic piezoelectric response of BFO. The unusual features of the response, i.e., a hard-to-soft transition induced by lowering the driving electric-field frequency and negative piezoelectric phase angle, can be explained by nonlinear piezoelectric Maxwell-Wagner effects. Simple analytical modeling confirms these macroscopic-response features and show that the key entities leading to such effects are conductive DWs. I could envision few points that can be drawn from the presented results. First, the data clearly show that local conductive paths (such as those along DWs) should be considered more seriously in addition to conventional bulk conductivity, which is mostly discussed in the literature on BFO. This is particularly important for the development of next-generation BFO-based piezoceramics for high-temperature applications as the local conductivity is what makes the response of BFO unstable in terms of driving field parameters (amplitude and frequency) and, most importantly, temperature. Second, the key problem related to M-W effects is that the response is boosted only at quasi-static driving conditions, as shown earlier by modeling of ceramic-polymer composites [45]. While certainly limited in the frequency range, it could be interesting to test anisotropic effects and piezoelectric enhancements in engineered matrices containing highly conducting charged DWs. It seems reasonable, though, that one should first validate the idea by modeling.

Acknowledgements

This work was financed by the Slovenian Research Agency in the frame of the program group P2-0105, projects J2-5483 and J2-9253, and PhD program of Maja Makarovic. I would like to thank Prof. Dragan Damjanovic for immense and fruitful discussions on domain walls and defect-related pinning effects as well as Maxwell-Wagner piezoelectric relaxation phenomena and modeling. Most of this chapter was written based on my knowledge that I received by working with Dragan. Further thanks go to the team at the Electronic Ceramics Department of the Jozef Stefan Institute for the constant support and never-ending interactive discussions. The research team is also acknowledged for extensive (micro)structural analyses and electric characterization of the carefully prepared samples over multiple scales of the material, without which complex mechanisms, such as those discussed in this chapter, cannot be seriously studied and thus explained.

Conflict of interest

The author of this work declares no conflict of interest of any kind.

Notes/thanks/other declarations

This chapter is dedicated to my wife Mihaela and my kids Aleks and Jan. Without them, my love (in the broadest possible physical and metaphysical sense) would have no meaning.

Author details

Tadej Rojac
Jozef Stefan Institute, Ljubljana, Slovenia

*Address all correspondence to: tadej.rojac@ijs.si

References

[1] Valasek J. Piezo-electric and allied phenomena in Rochelle salt. *Phys Rev* 1921; 17: 475–481.

[2] Brennecka G, Sherbondy R, Schwartz R, et al. Ferroelectricity - A revolutionary century of discovery. *Am Ceram Soc Bull* 2020; 99: 24–30.

[3] Scott JF, Paz De Araujo CA. Ferroelectric memories. *Science (80-)* 1989; 246: 1400–1405.

[4] Damjanovic D. Hysteresis in piezoelectric and ferroelectric materials. In: Mayergoyz I, Bertotti G (eds) *The Science of Hysteresis*. Elsevier, 2006, pp. 337–465.

[5] Hall DA. Nonlinearity in piezoelectric ceramics. *J Mater Sci* 2001; 36: 4575–4601.

[6] Bassiri-Gharb N, Fujii I, Hong E, et al. Domain wall contributions to the properties of piezoelectric thin films. *J Electroceramics* 2007; 19: 47–65.

[7] Cross LE. Ferroelectric ceramics: Tailoring properties for specific applications. In: Setter N, Colla EL (eds) *Ferroelectric ceramics*. Birkhäuser Basel, pp. 1–85.

[8] Lambeck PV, Jonker GH. The nature of domain stabilization in ferroelectric perovskites. *J Phys Chem Solids* 1986; 47: 453–461.

[9] Carl K, Hardtl KH. Electrical after-effects in $Pb(Zr,Ti)O_3$ ceramics. *Ferroelectrics* 1978; 17: 476–486.

[10] Jonker GH. Nature of aging in ferroelectric ceramics. *J Am Ceram Soc* 1972; 55: 57–58.

[11] Robels U, Arlt G. Domain wall clamping in ferroelectrics by orientation of defects. *J Appl Phys* 1993; 73: 3454–3460.

[12] Marincel DM, Zhang H, Jesse S, et al. Domain wall motion across various grain boundaries in ferroelectric thin films. *J Am Ceram Soc* 2015; 98: 1848–1857.

[13] Damjanovic D, Demartin M. Contribution of the irreversible displacement of domain walls to the piezoelectric effect in barium titanate and lead zirconate titanate ceramics. *J Phys Condens Matter* 1997; 9: 4943–4953.

[14] García JE, Ṕrez R, Ochoa DA, et al. Evaluation of domain wall motion in lead zirconate titanate ceramics by nonlinear response measurements. *J Appl Phys* 2008; 103: 054108.

[15] Eitel R, Randall CA. Octahedral tilt-suppression of ferroelectric domain wall dynamics and the associated piezoelectric activity in $Pb(Zr,Ti)O_3$. *Phys Rev B* 2007; 75: 094106.

[16] Lalitha K V., Riemer LM, Koruza J, et al. Hardening of electromechanical properties in piezoceramics using a composite approach. *Appl Phys Lett* 2017; 111: 022905.

[17] Arlt G, Neumann H. Internal bias in ferroelectric ceramics: Origin and time dependence. *Ferroelectrics* 1988; 87: 109–120.

[18] Ren X. Large electric-field-induced strain in ferroelectric crystals by point-defect-mediated reversible domain switching. *Nat Mater* 2004; 3: 91–94.

[19] Rojac T, Bencan A, Drazic G, et al. Domain-wall conduction in ferroelectric $BiFeO_3$ controlled by accumulation of charged defects. *Nat Mater* 2017; 16: 322–327.

[20] Mundy JA, Schaab J, Kumagai Y, et al. Functional electronic inversion layers at ferroelectric domain walls. *Nat Mater* 2017; 16: 622–627.

[21] Campanini M, Gradauskaite E, Trassin M, et al. Imaging and quantification of charged domain walls in $BiFeO_3$. *Nanoscale* 2020; 12: 9186–9193.

[22] Polking MJ. Deciphering the physics and chemistry of perovskites with transmission electron microscopy. *Nanoscale* 2016; 8: 6237–6248.

[23] Zhang QM, Pan WY, Jang SJ, et al. Domain wall excitations and their contributions to the weak-signal response of doped lead zirconate titanate ceramics. *J Appl Phys* 1988; 64: 6445–6451.

[24] Eitel RE, Shrout TR, Randall CA. Nonlinear contributions to the dielectric permittivity and converse piezoelectric coefficient in piezoelectric ceramics. *J Appl Phys* 2006; 99: 124110.

[25] Damjanovic D. Ferroelectric, dielectric and piezoelectric properties of ferroelectric thin films and ceramics. *Reports Prog Phys* 1998; 61: 1267–1324.

[26] Pramanick A, Damjanovic D, Daniels JE, et al. Origins of electro-mechanical coupling in polycrystalline ferroelectrics during subcoercive electrical loading. *J Am Ceram Soc* 2011; 94: 293–309.

[27] Damjanovic D. Stress and frequency dependence of the direct piezoelectric effect in ferroelectric ceramics. *J Appl Phys* 1997; 82: 1788–1797.

[28] Pramanick A, Damjanovic D, Nino JC, et al. Subcoercive cyclic electrical loading of lead zirconate titanate ceramics I: Nonlinearities and losses in the converse piezoelectric effect. *J Am Ceram Soc* 2009; 92: 2291–2299.

[29] Riemer LM, Jin L, Uršič H, et al. Dielectric and electro-mechanic nonlinearities in perovskite oxide ferroelectrics, relaxors, and relaxor ferroelectrics. *J Appl Phys* 2021; 129: 054101.

[30] Morozov M, Damjanovic D, Setter N. The nonlinearity and subswitching hysteresis in hard and soft PZT. *J Eur Ceram Soc* 2005; 25: 2483–2486.

[31] Damjanovic D, Demartin M. The Rayleigh law in piezoelectric ceramics. *J Phys D Appl Phys* 1996; 29: 2057–2060.

[32] Haugen AB, Morozov MI, Jones JL, et al. Rayleigh analysis of dielectric properties in textured $K_{0.5}Na_{0.5}NbO3$ ceramics. *J Appl Phys* 2014; 116: 214101.

[33] García JE, Guerra JDS, Arajo EB, et al. Domain wall contribution to dielectric and piezoelectric responses in $0.65Pb(Mg_{1/3}Nb_{2/3})O_3$-$0.35PbTiO_3$ ferroelectric ceramics. *J Phys D Appl Phys* 2009; 42: 115421.

[34] Otoničar M, Bradeško A, Fulanović L, et al. Connecting the multiscale structure with macroscopic response of relaxor ferroelectrics. *Adv Funct Mater* 2020; 30: 2006823.

[35] Griggio F, Jesse S, Kumar A, et al. Substrate clamping effects on irreversible domain wall dynamics in lead zirconate titanate thin films. *Phys Rev Lett* 2012; 108: 157604.

[36] Gharb NB, Trolier-McKinstry S, Damjanovic D. Piezoelectric nonlinearity in ferroelectric thin films. *J Appl Phys* 2006; 100: 044107.

[37] Robert G, Damjanovic D, Setter N. Preisach distribution function approach to piezoelectric nonlinearity and hysteresis. *J Appl Phys* 2001; 90: 2459–2464.

[38] García JE, Pérez R, Albareda A. Contribution of reversible processes to the non-linear dielectric response in hard lead zirconate titanate ceramics. *J Phys Condens Matter* 2005; 17: 7143–7150.

[39] Catalan G, Seidel J, Ramesh R, et al. Domain wall nanoelectronics. *Rev Mod Phys* 2012; 84: 119–156.

[40] Moreau JM, Michel C, Gerson R, et al. Ferroelectric $BiFeO_3$ x-ray and neutron diffraction study. *J Phys Chem Solids* 1971; 32: 1315–1320.

[41] Selbach SM, Tybell T, Einarsrud MA, et al. Phase transitions, electrical conductivity and chemical stability of $BiFeO_3$ at high temperatures. *J Solid State Chem* 2010; 183: 1205–1208.

[42] Stevenson T, Martin DG, Cowin PI, et al. Piezoelectric materials for high temperature transducers and actuators. *J Mater Sci Mater Electron* 2015; 26: 9256–9267.

[43] Dyre JC. The random free-energy barrier model for ac conduction in disordered solids. *J Appl Phys* 1988; 64: 2456–2468.

[44] Jonscher AK. *Dielectric relaxation in solids*. Chelsea Dielectric Press, 1983.

[45] Turik A V., Chernobabov AI, Radchenko GS, et al. Giant piezoelectric and dielectric enhancement in disordered heterogeneous systems. *Phys Solid State* 2004; 46: 2213–2216.

[46] Ueda H, Fukada E, Karasz FE. Piezoelectricity in three-phase systems: Effect of the boundary phase. *J Appl Phys* 1986; 60: 2672–2677.

[47] Furukawa T, Ishida K, Fukada E. Piezoelectric properties in the composite systems of polymers and PZT ceramics. *J Appl Phys* 1979; 50: 4904–4912.

[48] Damjanovic D, Demartin Maeder M, Duran Martin P, et al. Maxwell-Wagner piezoelectric relaxation in ferroelectric heterostructures. *J Appl Phys* 2001; 90: 5708–5712.

[49] Nguyen BH, Zhuang X, Rabczuk T. Numerical model for the characterization of Maxwell-Wagner relaxation in piezoelectric and flexoelectric composite material. *Comput Struct* 2018; 208: 75–91.

[50] Lunkenheimer P, Bobnar V, Bobnar V, et al. Origin of apparent colossal dielectric constants. *Phys Rev B* 2002; 66: 052105.

[51] Turik A V., Radchenko GS. Maxwell-Wagner relaxation in piezoactive media. *J Phys D Appl Phys* 2002; 35: 1188–1192.

[52] Turik A V., Radchenko GS. Maxwell-Wagner relaxation of elastic constants of layered polar dielectrics. *Phys Solid State* 2003; 45: 1060–1064.

[53] Hamano K, Yamaguchi T. Piezoelectric relaxation in ferroelectrics and polymers. *Ferroelectrics* 1982; 42: 23–33.

[54] Jones JL, Aksel E, Tutuncu G, et al. Domain wall and interphase boundary motion in a two-phase morphotropic phase boundary ferroelectric: Frequency dispersion and contribution to piezoelectric and dielectric properties. *Phys Rev B* 2012; 86: 024104.

[55] Gaunt P. Ferromagnetic domain wall pinning by a random array of inhomogeneities. *Philos Mag B* 1983; 48: 261–276.

[56] Lines ME, Glass MA. *Principles and applications of ferroelectric and related materials*. Clarendon Press, 1977.

[57] Wade RH. The determination of domain wall thickness in ferromagnetic films by electron microscopy. *Proc Phys Soc* 1962; 79: 1237–1244.

[58] Shilo D, Ravichandran G, Bhattacharya K. Investigation of twin-wall structure at the nanometre scale using atomic force microscopy. *Nat Mater* 2004; 3: 453–457.

[59] Blank H, Amelinckx S. Direct observation of ferroelectric domains in barium titanate by means of the electron microscope. *Appl Phys Lett* 1963; 2: 140–142.

[60] Morozov MI, Damjanovic D. Hardening-softening transition in Fe-doped $Pb(Zr,Ti)O_3$ ceramics and evolution of the third harmonic of the polarization response. *J Appl Phys* 2008; 104: 034107.

[61] Rojac T, Kosec M, Budic B, et al. Strong ferroelectric domain-wall pinning in $BiFeO_3$ ceramics. *J Appl Phys* 2010; 108: 074107.

[62] Rojac T, Drnovsek S, Bencan A, et al. Role of charged defects on the electrical and electromechanical properties of rhombohedral $Pb(Zr,Ti)O_3$ with oxygen octahedra tilts. *Phys Rev B* 2016; 93: 014102.

[63] Genenko YA, Glaum J, Hoffmann MJ, et al. Mechanisms of aging and fatigue in ferroelectrics. *Mater Sci Eng B* 2015; 192: 52–82.

[64] Erhart P, Träskelin P, Albe K. Formation and switching of defect dipoles in acceptor-doped lead titanate: A kinetic model based on first-principles calculations. *Phys Rev B* 2013; 88: 024107.

[65] Eichel RA. Structural and dynamic properties of oxygen vacancies in perovskite oxides - Analysis of defect chemistry by modern multi-frequency and pulsed EPR techniques. *Phys Chem Chem Phys* 2011; 13: 368–384.

[66] Zhang L, Erdem E, Ren X, et al. Reorientation of $(Mn_{Ti}''\text{-}V_O^{\bullet\bullet})^{\times}$ defect dipoles in acceptor-modified $BaTiO_3$ single crystals: An electron paramagnetic resonance study. *Appl Phys Lett* 2008; 93: 202901.

[67] Warren WL, Vanheusden K, Dimos D, et al. Oxygen vacancy motion in perovskite oxides. *Journal of the American Ceramic Society* 1996; 79: 536–538.

[68] Eichel RA, Erhart P, Träskelin P, et al. Defect-dipole formation in copper-doped $PbTiO_3$ ferroelectrics. *Phys Rev Lett* 2008; 100: 095504.

[69] Meštrić H, Eichel RA, Dinse KP, et al. Iron-oxygen vacancy defect association in polycrystalline iron-modified $PbZrO_3$ antiferroelectrics: Multifrequency electron paramagnetic resonance and Newman superposition model analysis. *Phys Rev B* 2006; 73: 184105.

[70] Eichel RA, Erünal E, Jakes P, et al. Interactions of defect complexes and domain walls in CuO-doped ferroelectric $(K,Na)NbO_3$. *Appl Phys Lett* 2013; 102: 242908.

[71] Aksel E, Erdem E, Jakes P, et al. Defect structure and materials 'hardening' in Fe_2O_3-doped $[Bi_{0.5}Na_{0.5}]TiO_3$ ferroelectrics. *Appl Phys Lett* 2010; 97: 012903.

[72] Postnikov VS, Pavlov VS, Turkov SK. Internal friction in ferroelectrics due to interaction of domain boundaries and point defects. *J Phys Chem Solids* 1970; 31: 1785–1791.

[73] Warren WL, Dimos D, Tuttle BA, et al. Electronic domain pinning in Pb(Zr, Ti)O_3 thin films and its role in fatigue. *Appl Phys Lett* 1994; 65: 1018–1020.

[74] Zhang L, Ren X. Aging behavior in single-domain Mn-doped $BaTiO_3$ crystals: Implication for a unified microscopic explanation of ferroelectric aging. *Phys Rev B* 2006; 73: 094121.

[75] Bencan A, Drazic G, Ursic H, et al. Domain-wall pinning and defect ordering in $BiFeO_3$ probed on the atomic and nanoscale. *Nat Commun* 2020; 11: 1762.

[76] Genenko YA, Lupascu DC. Drift of charged defects in local fields as aging mechanism in ferroelectrics. *Phys Rev B* 2007; 75: 184107.

[77] Genenko YA, Glaum J, Hirsch O, et al. Aging of poled ferroelectric ceramics due to relaxation of random depolarization fields by space-charge accumulation near grain boundaries. *Phys Rev B* 2009; 80: 224109.

[78] Chandrasekaran A, Damjanovic D, Setter N, et al. Defect ordering and defect-domain-wall interactions in $PbTiO_3$: A first-principles study. *Phys Rev B* 2013; 88: 214116.

[79] Rojac T, Ursic H, Bencan A, et al. Mobile domain walls as a bridge between nanoscale conductivity and macroscopic electromechanical response. *Adv Funct Mater* 2015; 25: 2099–2108.

[80] Makarovic M, Kanas N, Zorko A, et al. Tailoring the electrical conductivity and hardening in $BiFeO_3$ ceramics. *J Eur Ceram Soc* 2020; 40: 5483–5493.

[81] Eichel RA, Meštrić H, Kungl H, et al. Multifrequency electron paramagnetic resonance analysis of polycrystalline gadolinium-doped $PbTiO_3$ - Charge compensation and site of incorporation. *Appl Phys Lett* 2006; 88: 122506.

[82] Bernardo MS, Jardiel T, Peiteado M, et al. Intrinsic compositional inhomogeneities in bulk Ti-doped $BiFeO_3$: Microstructure development and multiferroic properties. *Chem Mater* 2013; 25: 1533–1541.

[83] Slouka C, Kainz T, Navickas E, et al. The effect of acceptor and donor doping on oxygen vacancy concentrations in lead zirconate titanate (PZT). *Materials (Basel)* 2016; 9: 945.

[84] Liu L, Rojac T, Damjanovic D, et al. Frequency-dependent decoupling of domain-wall motion and lattice strain in bismuth ferrite. *Nat Commun* 2018; 9: 4928.

[85] Damjanovic D. Logarithmic frequency dependence of the piezoelectric effect due to pinning of ferroelectric-ferroelastic domain walls. *Phys Rev B* 1997; 55: R649–R652.

[86] Georgiou V, Veksler D, Campbell JP, et al. Ferroelectricity in Polar Polymer-Based FETs: A Hysteresis Analysis. *Adv Funct Mater* 2018; 28: 1705250.

[87] Hayashi T, Yoshihara M, Ohmi SI, et al. Electrical properties of ferroelectric $BaMgF_4$ films grown on GaAs substrates using AlGaAs buffer layer. *Appl Surf Sci* 1997; 117–118: 418–422.

[88] Moulson AJ, Herbert JM. *Electroceramics*. Wiley & Sons, 2003.

[89] Uchino K, Hirose S. Loss mechanisms in piezoelectrics: How to measure different losses separately. *IEEE Trans Ultrason Ferroelectr Freq Control* 2001; 48: 307–321.

[90] Holland R. Representation of dielectric, elastic, and piezoelectric

losses by complex coefficients. *IEEE Trans Sonics Ultrason* 1967; 14: 18–20.

[91] Rojac T, Bencan A, Drazic G, et al. Piezoelectric nonlinearity and frequency dispersion of the direct piezoelectric response of $BiFeO_3$ ceramics. *J Appl Phys* 2012; 112: 064114.

[92] Seidel J, Martin LW, He Q, et al. Conduction at domain walls in oxide multiferroics. *Nat Mater* 2009; 8: 229–234.

[93] Condurache O, Dražić G, Sakamoto N, et al. Atomically resolved structure of step-like uncharged and charged domain walls in polycrystalline $BiFeO_3$. *J Appl Phys* 2021; 129: 054102.

[94] Makarovic M, Bayir MÇ, Ursic H, et al. Domain wall conductivity as the origin of enhanced domain wall dynamics in polycrystalline $BiFeO_3$. *J Appl Phys* 2020; 128: 064104.

[95] Rojac T, Makarovic M, Walker J, et al. Piezoelectric response of $BiFeO_3$ ceramics at elevated temperatures. *Appl Phys Lett* 2016; 109: 042904.

[96] Padurariu L, Curecheriu LP, Mitoseriu L. Nonlinear dielectric properties of paraelectric-dielectric composites described by a 3D Finite Element Method based on Landau-Devonshire theory. *Acta Mater* 2016; 103: 724–734.

[97] Daniel L, Hubert O, Billardon R. Homogenisation of magneto-elastic behaviour: From the grain to the macro scale. *Comput Appl Math* 2004; 23: 285–308.

[98] Daniel L, Rekik M, Hubert O. A multiscale model for magneto-elastic behaviour including hysteresis effects. *Arch Appl Mech* 2014; 84: 1307–1323.

[99] Labusch M, Keip MA, Shvartsman V V., et al. On the influence of ferroelectric polarization states on the magneto-electric coupling in two-phase composites. *Tech Mech* 2016; 36: 73–87.

[100] Sluka T, Tagantsev AK, Bednyakov P, et al. Free-electron gas at charged domain walls in insulating $BaTiO_3$. *Nat Commun* 2013; 4: 1808.

Section 3

Overall Design of the Actuators

Chapter 3

Bionic Type Piezoelectric Actuators

Shupeng Wang and Jianping Li

Abstract

Piezoelectric actuators have been applied in many research and industrial fields. However, how to improve the working performance of piezoelectric actuators is still a hot issue. Up to now, many new motion principles have been developed for new piezoelectric actuators. The bionic type piezoelectric actuator is a kind of the novel piezoelectric actuators, and it imitates the motion style of different creatures in nature to overcome the limitation of traditional piezoelectric actuators. Bionic type piezoelectric actuators are able to achieve large working stroke or large output force, which is of great significance for the development of piezoelectric actuators. The principle, design, and future of some bionic type piezoelectric actuators are discussed in this chapter.

Keywords: bionic piezoelectric actuator, inchworm actuator, seal actuator, stepping, long-range

1. Introduction

With the rapid development of science and engineering, ultra-precision positioning technology has been gradually becoming a common supporting technology in many fields such as integrated circuit, optical engineering, high-end manufacturing, biomedicine, and MEMS [1–3]. The actuator, which plays a big role, is one of the key parts in the ultra-precision positioning system. The stable output with millimeter-scale stroke and nanometer-scale resolution is the basic capacity for the actuator. Besides, other characteristics, such as quick response, wide speed range, and large loading capacity, are also important [4, 5]. The traditional actuators, such as stepping motors, hydraulics, and pneumatics equipment, have lots of advantages, including adequate positioning range, high stiffness, and large load capacity. Nevertheless, some performance defects, such as cumulative positioning error, wind up, and lost motion, cannot be eliminated [6, 7].

In order to acquire better performance of the actuators, some smart functional materials are developed as the actuating materials, such as piezoelectric materials [8], shape memory alloys [9], magnetostrictive materials [10], electrostrictive materials [11], and photostrictive materials [12]. Compared with other ones, piezoelectric materials possess many structural and functional advantages, such as large stiffness, compact size, quick response, high resolution, powerful output, and easy control. Therefore, they have been utilized in ultra-precision positioning systems more widely [13, 14]. Piezoelectric materials are crystals that have no inversion symmetry structures. Under external electric field, they can generate deformations because of the rotation of the internal electric domain by the inverse piezoelectric effect. All dielectric materials generate an electrostriction effect, but only crystal structures with no inversion symmetry can produce piezoelectric effect [12]. For electrostrictive materials, the relationship between the deformation and the electric field is parabolic, while for piezoelectric materials, it is linear [12]. Moreover, under different signal voltages, the piezoelectric materials can generate reversible expanding, contracting, and rotating deformations in one component [15]. Although the piezoelectric materials possess so many excellent characteristics, it is difficult to overcome a defect that the deformation of the piezoelectric materials is small [4, 5]. Due to the above defect, it is difficult to make use of the strain of the piezoelectric materials under the external electric field in the engineering world [16, 17]. Therefore, it has become a hot issue to develop piezoelectric actuators with a long work range and other excellent performance.

To make the piezoelectric actuator produce a long work range, researchers from all over the world propose a great many principles. The first principle is that a number of single layer piezoelectric components are stacked to form one multilayer piezoelectric actuator which can be called piezo-stack actuator. Using this method, many small deformations from the single-layer piezoelectric components can be concatenated to produce a long displacement of the multilayer piezoelectric actuator [18, 19]. The working range of a piezo-stack actuator can reach 0.1% to 0.15% of its dimension [20, 21]. The second principle is to utilize some designed mechanisms to enlarge the small deformation of the piezoelectric materials, such as lever mechanisms and polygon mechanisms. Using these enlarging mechanisms, we can obtain the submillimeter scale work range of the piezoelectric actuator [22, 23]. The third principle is the stepping principle which imitates some animals' movement behavior to repeat and accumulate numerous small displacements of the piezoelectric materials until an adequate stroke is achieved [24–27]. Many stepping-type piezoelectric actuators are bionic type piezoelectric actuators. The fourth principle is the ultrasonic driving principle which uses the high-frequency vibrations of the piezoelectric materials to drive the output component to produce large displacements [28–30].

This chapter introduces the working principle of the piezoelectric actuators with a long work range, especially the bionic type piezoelectric actuators. The actuators can

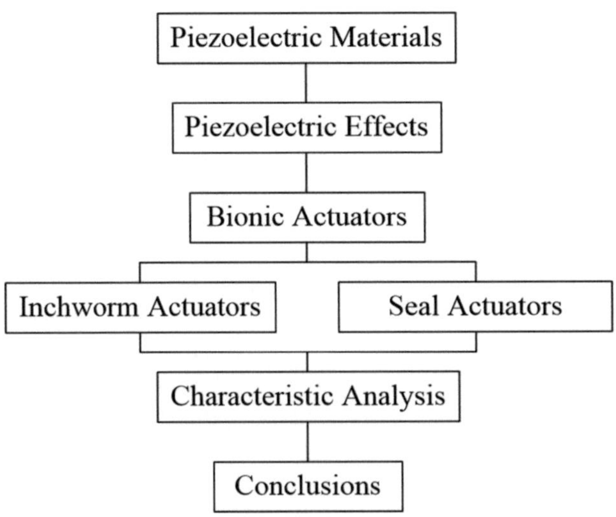

Figure 1.
The chapter organization.

be classified as linear ones and rotary ones. This chapter is mainly elaborated from the linear actuators. But the theories in this chapter are also suitable for the rotary actuators. The organization of the chapter is shown in **Figure 1**. The chapter is designed as follows: In Section 2, the piezoelectric materials and the piezoelectric effects are introduced briefly. In Section 3, the inchworm type actuators are elaborated, which are classified into the walker type, the pusher type, and the mixed type. In Section 4, we describe the seal type actuators including the walker type, the pusher type, and the mixed type. In Section 5, the characteristics of the bionic actuators are analyzed. In the last section, the conclusions of the chapter are given.

2. Piezoelectric effects

Piezoelectric materials are kind of crystals that have no inversion symmetry structures. As is known to all, under an external electric field, piezoelectric materials can generate deformations because of the rotation of the internal electric domain. When it comes to piezoelectric materials, piezoelectric effects are the most special functional characteristics including the direct piezoelectric effect and the converse piezoelectric effect [31–34]. When the mechanical stress is applied to the piezoelectric materials, electric charges can be produced on the electrodes of the piezoelectric materials, which is called the direct piezoelectric effect. Conversely, when electric voltages are applied on the electrodes of the piezoelectric materials, mechanical deformations can be generated on the piezoelectric materials, which is called the converse piezoelectric effect [33, 34]. In a real application, using the direct piezoelectric effect, piezoelectric materials can be employed as sensors. Using the converse piezoelectric effect, piezoelectric materials can be employed as actuators [33, 34].

PZT ($Pb(Zr_xTi_{1-x})O_3$) is a kind of piezoelectric ceramic material with excellent performance which has been used widely. Usually, PZT can be manufactured to be sheet, circular, ring, block, and so on. For the piezoelectric materials, there are three key parameters influencing the piezoelectric effects: the output strain δ, polarization field P, and actuation field E. According to the orientations of the three key parameters, the piezoelectric actuators can function in three working modes: longitudinal mode, transversal mode, and shear mode [35, 36]. As **Figure 2(a)** shows, in the manufacturing process, a block of piezoelectric material is polarized and the polarization field is P. The coordinate system on the piezoelectric material is established and the six-coordinate axes are named x (1), y (2), z (3), θ_x (4), θ_y (5) and θ_z

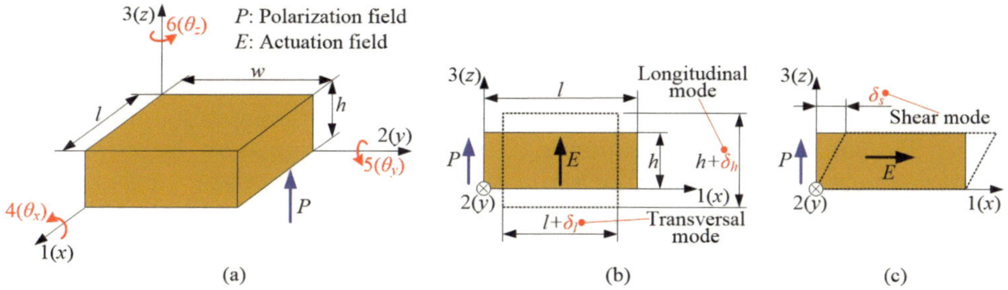

Figure 2.
Converse piezoelectric effect [32]. (a) Actuation and polarization fields; (b) longitudinal and transversal modes; (c) shear mode.

(6), respectively. The electric field E is applied to actuate the piezoelectric material. If the actuation field E and the polarization field P have the same direction or the opposite direction, deformations from the longitudinal direction and the transversal direction are generated simultaneously, which are named δ_h and δ_l, respectively. Correspondingly, these are the longitudinal mode and transversal mode of the piezoelectric material, respectively (see **Figure 2(b)**). If the actuation field E and the polarization field P have the vertical directions, deformation from the shear direction is generated, which is named δ_s. This is the shear mode of the piezoelectric material (see **Figure 2(c)**) [35, 36].

3. Inchworm actuators

The inchworm is a kind of insect in nature, which moves by wriggle. The researchers found that the inchworm has special body structures including a flexible body and numerous feet. The flexible body can be bent and the feet are on the head end and the tail end of the body. When the inchworm moves, it grips the tree trunk firstly with its forefeet on the head end of the body. Next, the inchworm bends its flexible body and pulls forward the hindfeet on the tail end of the body. Then, it grips the tree trunk with its hindfeet. Afterward, the inchworm releases the forefeet from the tree trunk. Then, it straightens the flexible body and pushes forward the forefeet. Finally, the inchworm grips the tree trunk again with its forefeet and releases the hindfeet from the tree trunk. After the above actions, the inchworm has moved forward with one step and reverts to the initial status. If the above process is repeated continuously, the inchworm can move forward continuously on the tree trunk.

Inspired by the wriggle mode of the inchworm, the researchers develop the inchworm type actuators driven by the piezoelectric materials. Similarity with the body structures of the inchworm, the inchworm type piezoelectric actuator is composed of a feeding component and two clamping components, which imitate the flexible body, forefeet, and hindfeet of the inchworm respectively. As shown in **Figure 3**, according to the relative position relationship of the feeding component and two clamping components, the inchworm actuator is classified as walker type, pusher type, and mixed type. As shown in **Figure 3(a)**, the walker type actuator is whose feeding component and two clamping components are all designed on the mover. During the process of the walker type actuator running, all of the feeding component and two clamping components are pushed with the mover. As shown in **Figure 3(b)**, the pusher type actuator is whose feeding component and two clamping components are all designed on the stator. During the process of the pusher type actuator running, none of the feeding component and two clamping components is pushed with the mover. As shown in **Figure 3(c)**, the mixed type actuator combines the structural features of the walker and pusher piezoelectric actuators. Either the feeding component or the two clamping components is designed on the mover and the other component is designed on the stator. During the process of the pusher type actuator running, Either the feeding component or the two clamping components is pushed with the mover and the other component is fixed on the stator.

The operating principles for the three types of inchworm actuators are shown in **Figure 3** and introduced as follows:

a. The operating principle of the walker type actuator:

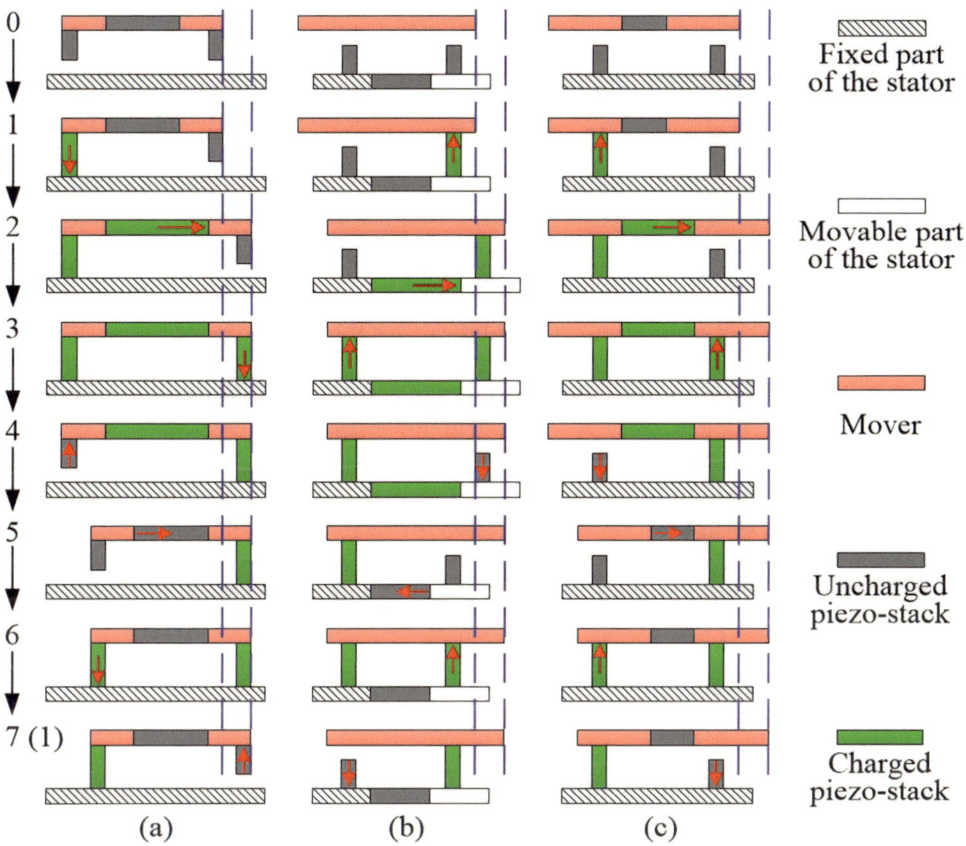

Figure 3.
Operating principles of inchworm actuators [32]. (a) Walker type; (b) pusher type; (c) mixed type.

1. The initial status of the walker actuator;
2. The left clamping component operates and clamps the left end of the stator;
3. The feeding component functions and pushes the right end of the mover rightwards;
4. The right clamping component operates and clamps the right end of the stator;
5. The left clamping component resets and releases the left end of the stator;
6. The feeding component resets and pulls the left end of the mover rightwards;
7. The left clamping component operates and clamps the left end of the stator again;
8. The right clamping component resets and releases the right end of the stator. The actuator reverts to step (1).

After the above actions, the mover of the walker actuator has moved rightwards with one step. If steps (1) to (7) are repeated continuously, the walker inchworm actuator can move rightwards to output long-range displacement step by step. The backward motion can be generated if the operating sequences of feeding component and the clamping components are changed.

b. The operating principle of the pusher type actuator:

1. The initial status of the pusher actuator;
2. The right clamping component operates and clamps the right end of the mover;
3. The feeding component functions and pushes the mover rightwards;
4. The left clamping component operates and clamps the left end of the mover;
5. The right clamping component resets and releases the right end of the mover;
6. The feeding component resets and pulls the right end of the stator leftwards;
7. The right clamping component operates and clamps the right end of the mover again;
8. The left clamping component resets and releases the left end of the mover. The actuator reverts to step (1).

After the above actions, the mover of the pusher actuator has moved rightwards with one step. If steps (1) to (7) are repeated continuously, the pusher inchworm actuator can move rightwards to output long-range displacement step by step. The backward motion can be generated if the operating sequences of feeding component and the clamping components are changed.

c. The operating principle of the mixed type actuator:

1. The initial status of the mixed actuator;
2. The left clamping component operates and clamps the left end of the mover;
3. The feeding component functions and pushes the right end of the mover rightwards;
4. The right clamping component operates and clamps the right end of the mover;
5. The left clamping component resets and releases the left end of the mover;
6. The feeding component resets and pulls the left end of the mover rightwards;
7. The left clamping component operates and clamps the left end of the mover again;
8. The right clamping component resets and releases the right end of the mover. The actuator reverts to step (1).

After the above actions, the mover of the mixed actuator has moved rightwards with one step. If steps (1) to (7) are repeated continuously, the mixed inchworm actuator can move rightwards to output long-range displacement step by step. The backward motion can be generated if the operating sequences of feeding component and the clamping components are changed. Generally, the electric signals to drive the inchworm actuator are usually the rectangular wave signal or trapezoidal wave signal.

4. Seal actuators

The seal is a kind of mammal living in the ocean. It is found that the seal has a flexible body with the contracting ability, two forefeet with the walking ability, and two hindfeet losing the walking ability. When the seal moves, it grips the sand beach firstly with its forefeet. Next, the seal contracts its flexible body and pulls forward the tail and the hindfeet. Then, it releases the forefeet from the sand beach. Afterwards, the seal stretches the flexible body and pushes forward the head and the forefeet. Finally, the seal grips the sand beach again with its forefeet. After the above actions, the seal has moved forward with one step and reverts to the initial status. If the above process is repeated continuously, the seal can move forward continuously on the sand beach.

Inspired by the wriggle mode of the seal, the researchers develop the seal-type actuators driven by piezoelectric materials. The inchworm type actuator consists of two clamping components that are both intermittent clamping mechanisms that can generate intermittent clamping force as required. If one of the two intermittent type clamping components is replaced by a persistent clamping component that can only generate constant clamping force, the inchworm type actuator will be changed into a seal type actuator. Similarity with the body structures of the seal, the seal type piezoelectric actuator is composed of a feeding component, an intermittent clamping component, and a persistent clamping component, which imitate the flexible body, the forefeet, and the hindfeet of the seal respectively.

As shown in **Figure 4**, according to the relative position relationship of the feeding component, the intermittent clamping component, and the persistent clamping component, the seal actuator is classified to walker type, pusher type, and mixed type. As shown in **Figure 4(a)**, the walker type actuator is whose feeding component, intermittent clamping component and persistent clamping component are all designed on the mover. During the process of the walker type actuator running, all of the feeding component, the intermittent clamping component, and the persistent clamping component are pushed with the mover. As shown in **Figure 4(b)**, the pusher type actuator is whose feeding component, intermittent clamping component and persistent clamping component are all designed on the stator. During the process of the pusher type actuator running, none of the feeding component, the intermittent clamping component, and the persistent clamping component is pushed with the mover. As shown in **Figure 4(c)**, the mixed type actuator combines the structural features of the walker and pusher piezoelectric actuators. One of the feeding component, the intermittent clamping component and the persistent clamping component is designed on the mover, and the other components are designed on the stator. During the process of the pusher type actuator running, one of the feeding component, the intermittent clamping component and the persistent clamping component is pushed with the mover and the other components are fixed on the stator.

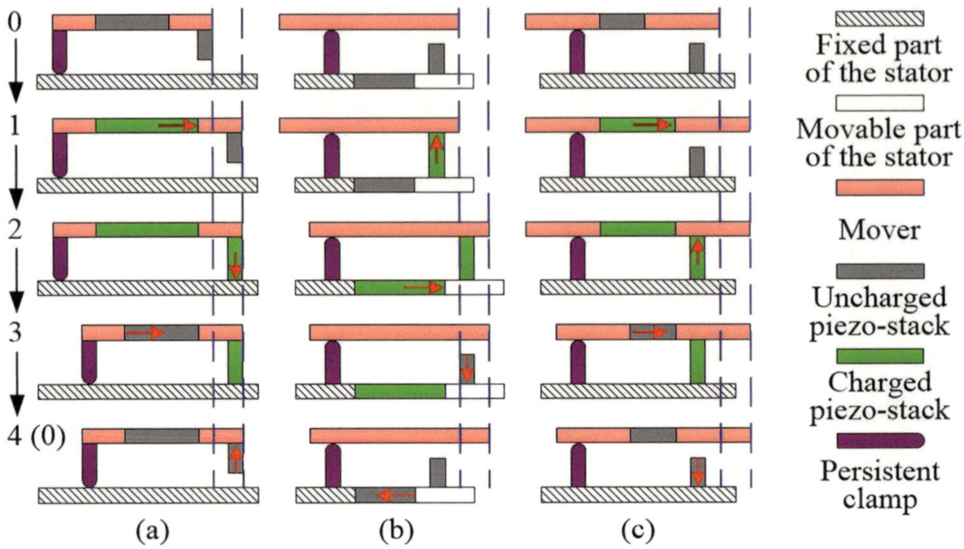

Figure 4.
Operating principles of seal actuators [32]. (a) Walker type; (b) pusher type; (c) mixed type.

The operating principles for the three types of seal actuators are shown in **Figure 4** and introduced as follows:

a. The operating principle of the walker type actuator:

1. The initial status of the walker actuator;

2. The feeding component functions and pushes rightwards the intermittent clamping component on the right side of the mover;

3. The intermittent clamping component on the right side of the mover operates and clamps the right side of the stator;

4. The feeding component resets and pulls rightwards the persistent clamping component on the left side of the mover;

5. The intermittent clamping component on the right side of the mover resets and releases from the right side of the stator. The actuator reverts to step (0).

After the above actions, the mover of the walker actuator has moved rightwards with one step. If the steps (0) to (4) are repeated continuously, the walker seal actuator can move rightwards to output long-range displacement step by step. The backward motion can be generated if the operating sequences of feeding component and the intermittent clamping component are changed.

b. The operating principle of the pusher type actuator:

1. The initial status of the pusher actuator;

2. The intermittent clamping component on the right side of the stator operates and clamps the right side of the mover;

3. The feeding component functions and pushes rightwards the mover;

4. The intermittent clamping component on the right side of the stator resets and releases from the right side of the mover;

5. The feeding component resets and pulls leftwards the intermittent clamping component on the right side of the stator. The actuator reverts to step (0).

After the above actions, the mover of the pusher actuator has moved rightwards with one step. If the steps (0) to (4) are repeated continuously, the pusher seal actuator can move rightwards to output long-range displacement step by step. The backward motion can be generated if the operating sequences of feeding component and the intermittent clamping component are changed.

c. The operating principle of the mixed type actuator:

1. The initial status of the mixed actuator;

2. The feeding component functions and pushes rightwards the right side of the mover;

3. The intermittent clamping component on the right side of the stator operates and clamps the right side of the mover;

4. The feeding component resets and pulls rightwards the left side of the mover;

5. The intermittent clamping component on the right side of the stator resets and releases from the right side of the mover. The actuator reverts to step (0).

After the above actions, the mover of the mixed actuator has moved rightwards with one step. If the steps (0) to (4) are repeated continuously, the mixed seal actuator can move rightwards to output long-range displacement step by step. The backward motion can be generated if the operating sequences of feeding component and the intermittent clamping component are changed. Generally, the electric signals to drive the seal actuator are usually the rectangular wave signal, the trapezoidal wave signal or the triangular wave signal.

5. Characteristic analysis of the bionic actuators

The inchworm actuator and the seal actuator are the commonest bionic type piezoelectric actuators, which are also the stepping type actuators. With the stepping driving principles, the piezoelectric actuators are completely free of the micro deformations of the piezoelectric materials and are able to generate the long-range displacements as required step by step. Therefore, the output displacements can reach numerous millimeters, or even without limit. The characteristics of the two types of bionic actuators are presented in **Table 1**.

Item\Type	Inchworm	Seal
Structure	Complex	Compact
Resolution	Medium	High
Capacity	large	Small
Velocity	Low	High
Control	Complex	Simple

Table 1.
Characteristics of the bionic actuators.

The inchworm actuator has a complex configuration while the structure of the seal actuator is compact. Thus, if there are requirements for the space and the weight, the seal actuator is preferred. Generally, the resolution of the seal actuator is higher than that of the inchworm. Hence, we use the seal actuator a lot to achieve more precise operation. The inchworm actuator has a larger load capacity than the seal actuator, so we can select the inchworm actuator when we need the large output force. If high speed is required, the seal actuator is more appropriate. Because the seal actuator is able to generate higher velocity than the inchworm actuator. In addition, the seal actuator is easier to control than the inchworm actuator. Because the inchworm actuator has more moving executive components. In application, we can select and use an appropriate type of bionic actuators according to **Table 1**.

6. Conclusions

In this chapter, the bionic type piezoelectric actuators with long-range outputs are introduced. Firstly, we present the frequently-used piezoelectric materials and the piezoelectric effects including the direct piezoelectric effect and the converse piezoelectric effect. Next, the inchworm type actuators are elaborated, which are classified into the walker type, the pusher type, and the mixed type. Then, we describe the seal type actuators including the walker type, the pusher type, and the mixed type. Afterwards, the characteristics of the bionic actuators are discussed. The configurations, classifications, principles, connections, and distinctions of the bionic type actuators are all presented in the chapter. We not only analyze the advantages and disadvantages for each type, but also discuss the derivation relationships among the actuators. This chapter conduces to readers to study the bionic piezoelectric actuators and is conducive to accomplish effective designs and future breakthroughs in technology.

Acknowledgements

This work was supported by the Talent Introduction Fund of Jilin University under Grant 451210330007.

Conflict of interest

The authors declare no conflict of interest.

Author details

Shupeng Wang[1] and Jianping Li[2]*

1 Jilin University, Changchun, China

2 Zhejiang Normal University, Jinhua, China

*Address all correspondence to: lijp@zjnu.cn

References

[1] Cheng L, Liu W, Hou ZG, et al. Neural-network-based nonlinear model predictive control for piezoelectric actuators. IEEE Transactions on Industrial Electronics. 2015;**62**(12):7717-7727. DOI: 10.1109/tie.2015.2455026

[2] Lin FJ, Hung YC, Chen SY. FPGA-based computed force control system using Elman neural network for linear ultrasonic motor. IEEE Transactions on Industrial Electronics. 2009;**56**(4):1238-1253. DOI: 10.1109/TIE.2008.2007040

[3] Lee SW, Ahn KG, Ni J. Development of a piezoelectric multi-axis stage based on stick-and-clamping actuation technology. Smart Materials and Structures. 2007;**16**(6):2354-2367. DOI: 10.1088/0964-1726/16/6/040

[4] Wang S, Rong W, Wang L, et al. Design, analysis and experimental performance of a bionic piezoelectric rotary actuator. Journal of Bionic Engineering. 2017;**14**(002):348-355. DOI: 10.1016/S1672-6529(16)60403-1

[5] Kang BW, Kim J. Design, fabrication, and evaluation of stepper motors based on the piezoelectric torsional actuator. IEEE/ASME Transactions on Mechatronics. 2013;**18**(6):1850-1854. DOI: 10.1109/TMECH.2013.2269171

[6] Henke B, Sawodny O, Neumann R. Distributed parameter Modeling of flexible ball screw drives using Ritz series discretization. IEEE/ASME Transactions on Mechatronics. 2015;**20**(3):1226-1235. DOI: 10.1109/TMECH.2014.2333775

[7] Wang S, Rong W, Wang L, et al. Design, analysis and experimental performance of a novel stick-slip type piezoelectric rotary actuator based on variable force couple driving. Smart Materials and Structures. 2017;**26**(5):055005. DOI: 10.1088/1361-665X/aa64c3

[8] Liu Y, Chen W, Yang X, et al. A rotary piezoelectric actuator using the third and fourth bending vibration modes. IEEE Transactions on Industrial Electronics. 2014;**61**(8):4366-4373. DOI: 10.1109/TIE.2013.2272279

[9] Hwang D, Ihn YS, Kim K. Compact modular cycloidal motor with embedded shape memory alloy wires. IEEE Transactions on Industrial Electronics. 2018;**PP**(5):4028-4038. DOI: 10.1109/TIE.2017.2764839

[10] Santapuri S, Scheidler JJ, Dapino MJ. Two-dimensional dynamic model for composite laminates with embedded magnetostrictive materials. Composite Structures. 2015;**132**(11):737-745. DOI: 10.1016/j.compstruct.2015.04.062

[11] Lallart M, Richard C, Sukwisut P, et al. Electrostrictive bending actuators: Modeling and experimental investigation. Sensors and Actuators, A: Physical. 2012;**179**:169-177. DOI: 10.1016/j.sna.2012.02.023

[12] Mirvakili SM, Hunter IW. Artificial muscles: Mechanisms, applications, and challenges. Advanced Materials. 2018;**30**(6):1704407. DOI: 10.1002/adma.201704407

[13] Akin YY, Toprak A, Tigli O. Piezoelectric membrane actuators for micropump applications using PVDF-TrFE. Journal of Microelectromechanical Systems. 2018;**27**:86-94. DOI: 10.1109/JMEMS.2017.2773090

[14] Lucinskis R, Mazeika D, Bansevicius R. Investigation of

oscillations of piezoelectric actuators with multi-directional polarization. Mechanical Systems and Signal Processing. 2018;**99**:450-458. DOI: 10.1016/j.ymssp.2017.06.036

[15] Morvan J, Buyuktanir E, West JL, et al. Highly piezoelectric biocompatible and soft composite Fibers. Applied Physics Letters. 2012;**100**(6):063901. DOI: 10.1063/1.3683482

[16] Li J, Zhou X, Zhao H, et al. Development of a novel parasitic-type piezoelectric actuator. IEEE/ASME Transactions on Mechatronics. 2017;**22**:541-550. DOI: 10.1109/TMECH.2016.2604242

[17] Poikselkä KL, Leinonen M, Palosaari J, Vallivaara I, Röning J, Juuti J. Novel genetically optimised high-displacement piezoelectric actuator with efficient use of active material. Smart Materials and Structures. 2017;**26**(9):095022. DOI: 10.1088/1361-665X/aa770a

[18] Kaufmann P, RöHrig S, Supancic P, et al. Influence of ferroelectric domain texture on the performance of multilayer piezoelectric actuators. Journal of the European Ceramic Society. 2017;**37**(5):2039-2046. DOI: 10.1016/j.jeurceramsoc.2016.12.029

[19] Esteves G, Fancher CM, Rhrig S, et al. Electric-field-induced structural changes in multi-layer piezoelectric actuators during electrical and mechanical loading. Acta Materialia. 2017;**132**:96-105. DOI: 10.1016/j.actamat.2017.04.014

[20] Ho ST, Jan SJ. A piezoelectric motor for precision positioning applications. Precision Engineering. 2016;**43**:285-293. DOI: 10.1016/j.precisioneng.2015.08.007

[21] Zhang Y, Zhang WJ, Hesselbach J, et al. Development of a two-degree-of-freedom piezoelectric rotary-linear actuator with high driving force and unlimited linear movement. Review of Scientific Instruments. 2006;**77**(3):465-481. DOI: 10.1063/1.2185500

[22] Nam J, Oh H, Jang G. Externally leveraged resonant piezoelectric actuator with fast response time for smart devices. IEEE/ASME Transactions on Mechatronics. 2016;**21**(6):2764-2772. DOI: 10.1109/TMECH.2016.2590564

[23] Wang S, Rong W, Wang L, et al. A novel inchworm type piezoelectric rotary actuator with large output torque: Design, analysis and experimental performance[J]. Precision Engineering. 2017;**51**:545-551. DOI: 10.1016/j.precisioneng.2017.10.010

[24] Salisbury SP, Waechter DF, Mrad RB, et al. Design considerations for complementary inchworm actuators. IEEE/ASME Transactions on Mechatronics. 2006;**11**(3):265-272. DOI: 10.1109/TMECH.2006.875565

[25] Belly C, Charon W. Benefits of amplification in an inertial stepping motor. Mechatronics. 2012;**22**:177-183. DOI: 10.1016/j.mechatronics.2012.01.006

[26] Yokozawa H, Morita T. Wireguide driving actuator using resonant-type smooth impact drive mechanism. Sensors and Actuators, A: Physical. 2015;**230**:40-44. DOI: 10.1016/j.sna.2015.04.012

[27] Li J, Huang H, Morita T. Stepping piezoelectric actuators with large working stroke for nano-positioning systems: A review. Sensors and Actuators A: Physical. 2019;**292**:39-51. DOI: 10.1016/j.sna.2019.04.006

[28] Pott PP, Matich S, Schlaak HF. Ultrasonic resonant piezoelectric actuator with intrinsic torque measurement.

IEEE Transactions on Ultrasonics, Ferroelectrics, and Frequency Control. 2012;**59**(11):2509-2514. DOI: 10.1109/TUFFC.2012.2483

[29] Chu X, Wang J, Yuan S, et al. A screw-thread-type ultrasonic actuator based on a Langevin piezoelectric vibrator. Review of Scientific Instruments. 2014;**85**(6):065002. DOI: 10.1063/1.4882975

[30] Marquez IAR, Bolborici V. A dynamic model of the piezoelectric traveling wave rotary ultrasonic motor stator with the finite volume method. Ultrasonics. 2017;**77**:69-78. DOI: 10.1016/j.ultras.2017.01.019

[31] Ma J, Hu Y, Li B, et al. Influence of secondary converse piezoelectric effect on deflection of fully covered PZT actuators. Sensors and Actuators, A: Physical. 2012;**175**:132-138. DOI: 10.1016/j.sna.2011.12.034

[32] Wang S, Rong W, Wang L, et al. A survey of piezoelectric actuators with long working stroke in recent years: Classifications, principles, connections and distinctions. Mechanical Systems and Signal Processing. 2019;**123**:591-605. DOI: 10.1016/j.ymssp.2019.01.033

[33] Liu Z, Yao Z, Li X, et al. Design and experiments of a linear piezoelectric motor driven by a single mode. Review of Scientific Instruments. 2016;**87**(11):273-768. DOI: 10.1063/1.4966251

[34] Zhu C, Chu X, Yuan S, et al. Development of an ultrasonic linear motor with ultra-positioning capability and four driving feet. Ultrasonics. 2016;**72**:66-72. DOI: 10.1016/j.ultras.2016.07.010

[35] Gibert JM, Austin EM. Demonstration of optimizing piezoelectric polarization in the design of a flextensional actuator. Structural and Multidisciplinary Optimization. 2007;**33**:471-480. DOI: 10.1007/s00158-006-0052-8

[36] Driesen W. Concept, modeling and experimental characterization of the modulated friction inertial drive (MFID) locomotion principle: Application to mobile microrobots [thesis]. École Polytechnique Fédérale de Lausanne. 2008

Chapter 4

Design, Analysis and Testing of Piezoelectric Tool Actuator for Elliptical Vibration Cutting

Jinguo Han, Mingming Lu and Jieqiong Lin

Abstract

In the field of ultraprecision machining, the structured surfaces with various micro/nano characteristics may have different advanced functions, such as wettability modifications, tribological control and hybrid micro-optics. However, the machining of micro/nano structured surfaces is becoming a challenge for present cutting method. Especially for the difficult-to-cut materials, it is impossible to manufacture complex micro/nano features by using traditional cutting methods. The complex features require a cutting tool no longer confined to the traditional motion guide. The cutting tool should have more quick response velocity and flexible modulated ability. This chapter aims to make an introduction for piezoelectric tool actuator used in elliptical vibration cutting, which can be offering tertiary cutting operations with quick response and flexible modulated ability. The content covers the working principle of piezoelectric tool actuator, compliant mechanism design, static modeling, kinematic and dynamic modeling, structure optimization and offline testing.

Keywords: piezoelectric actuator, elliptical vibration cutting, compliant mechanism, ultraprecision machining, structured surface

1. Introduction

With the rapid growth of science and technology, precision components are increasingly in demand in various fields such as aerospace, biomedical engineering, optics, surface engineering and energy. These components not only require tight tolerances and high-quality surface finishes but also require the use of difficult-to-cut materials like high-temperature alloys and hard-brittle materials owing to the physical, mechanical, optical and electronic properties [1]. It is because of the difficult-to-cut characteristics, the precision machining of these materials has always been a challenge. For traditional machining methods, there are always some disadvantages, such as high cutting temperature and cutting force, severe tool wear, low efficiency and poor surface quality. To improve the manufacturing quality, tool life and machining efficiency, let the machining cost down, elliptical vibration-assisted machining (EVAM) was proposed. In ultraprecision machining, especially the structured surface with micro-nano features, it requires exceptionally fine and repeatable motions, which makes the piezoelectric actuator a popular candidate for EVAM.

In EVAM, the elliptical vibration cutting (EVC) is a typical method for turning. It has been validated that EVC can improve the machinability of difficult-to-cut materials [2–4]. Meanwhile, the lower cutting force, longer tool life and better machining quality were obtained. In EVC, the vibration of cutting tool or workpiece is usually motivated by external excitation. For elliptical vibration cutting, the locus of cutting tool or workpiece is an ellipse due to the deliberate modulation of vibration devices. According to the principle of vibration, vibration devices can be divided into two types: resonant vibration and non-resonant vibration. For resonant type EVC vibrator, usually consists of an ultrasonic generator, a transducer and an ultrasonic vibration horn. The working frequency is generally above 20 kHz. There are two design schemes for the structure configuration, i.e. patch and sandwich. For the patch type EVC vibrator, two sets of piezoelectric plates are attached to the outer wall of the resonant rod to achieve the same or different modes of resonance. What's more, the Langevin-type transducer is usually adopted in the sandwich-type EVC vibrator. The vibrator can be achieved by adding two or more sets of piezoelectric rings to the resonant rod or coupling two Langevin-type transducers. The ellipse locus of cutting tool can be obtained through the same or different modes coupling in different directions with the nearly same vibration frequency. The working frequency is the resonance frequency. Additionally, the fixed working frequency and vibration parameters, poor dynamic accuracy and open-loop control, heat generation problems are the limitation problems for the resonant vibrator.

For non-resonant type EVC vibrator, usually consists of piezoelectric actuators and a compliant mechanism with deliberate design. Compared with the resonant vibration, the non-resonant type EVC vibrator provides much higher flexibility due to the elastic deformation of compliant mechanism. The vibration parameters like amplitude and frequency are easy to be controlled. The micro-nano motion resulting from elastic deformation of flexure hinge can guarantee the modulation ability and accuracy, which makes it more suitable for the manufacturing of micro-nano structured surfaces.

Generally, compared with workpiece vibration, tool vibration is more popular due to the ease of implementation. In fact, the modulation capability of tools plays a very important role in manufacturing for different materials and structure features. However, the tool modulation capability mainly depends on the performance of the vibrator. Obviously, the non-resonant type EVC vibrator has more advantages than the resonant type EVC vibrator. Up to now, there are many different types of non-resonant EVC vibrators were developed for vibration machining. Therefore, this chapter mainly focuses on the non-resonant type EVC vibrator, aiming to review the working principle of piezoelectric tool actuator, compliant mechanism design, static modeling, kinematic and dynamic modeling, structure optimization and offline testing.

2. Introduction to the different types of EVC

EVC is a kind of vibration-assisted machining. Different from 1-dimensional vibration cutting and traditional cutting, the trajectory of cutting tool is an ellipse and it is located in a fixed plane (2-dimensional, 2D) or in space (3-dimensional, 3D) which is dependent on the vibration modulation of motion axis. The cutting process of different types of EVC is shown in **Figure 1**. It can be seen that for orthogonal EVC and oblique EVC, the tool trajectory is in a fixed plane despite the orientation of fixed

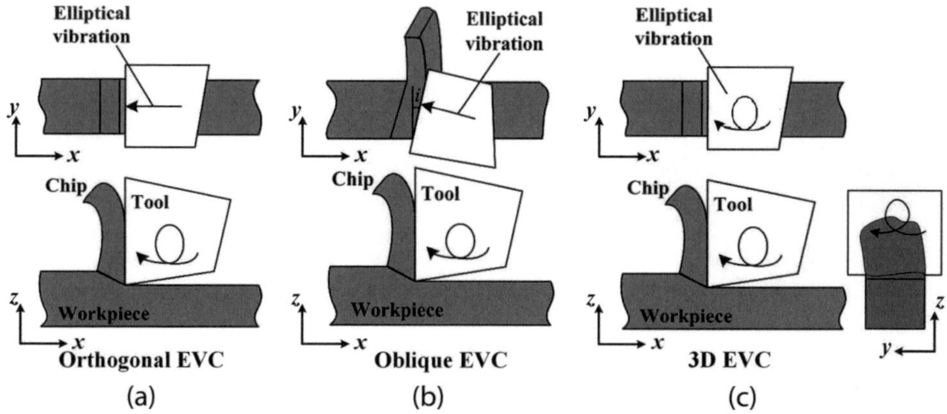

Figure 1.
Schematic illustration of different types of EVC.

plane in oblique EVC is also dependent on the angle i. It only needs two motion axes to modulate the cutting tool to obtain the ellipse locus. However, for 3D EVC, the trajectory of cutting tool is in space or in an arbitrary plane of modulation range. Therefore, it needs at least three motion axes to obtain the ellipse locus.

3. Development of the non-resonant type EVC vibrator

3.1 2D type EVC vibrator

The EVC method is first proposed based on a 2D type EVC vibrator. Therefore, the research on 2D type EVC vibrator is relatively early and extensive. According to the principle of EVC, generally, there are two piezoelectric actuator configurations for the simplest: one is there existing an angle between two piezoelectric actuators, the other one is piezoelectric actuator parallel configuration. Shamoto and Moriwaki applied the vertical piezoelectric actuator configuration to achieve the EVC [5]. Cerniway [6] and Negishi [7] applied parallel piezoelectric actuator configuration to achieve the EVC. However, these non-resonant EVC vibrators have limited modulation ability and severe crosstalk. In recent years, the flexure-based non-resonant EVC vibrator has been developing rapidly. Here are some typical examples following below.

3.1.1 A hybrid flexure hinge EVC vibrator actuated by parallel piezoelectric actuator configuration

3.1.1.1 Orthogonal type

In this work, a hybrid flexure hinge EVC vibrator actuated by parallel piezoelectric actuators was proposed [8]. The mechanical structure of the vibrator is shown in **Figure 2(a)**. The coordinate system O_{out}-xyz is fixed on the end-effector. The mechanical structure has mirror symmetry about the y-axis. Two piezoelectric actuators, which are placed in parallel from top to bottom and directly preloaded by two screws, are used to actuate the motion of the cutting tool. There are four motion guidance components in the mechanical structure, i.e. two double parallel four-bar linkage mechanisms and two

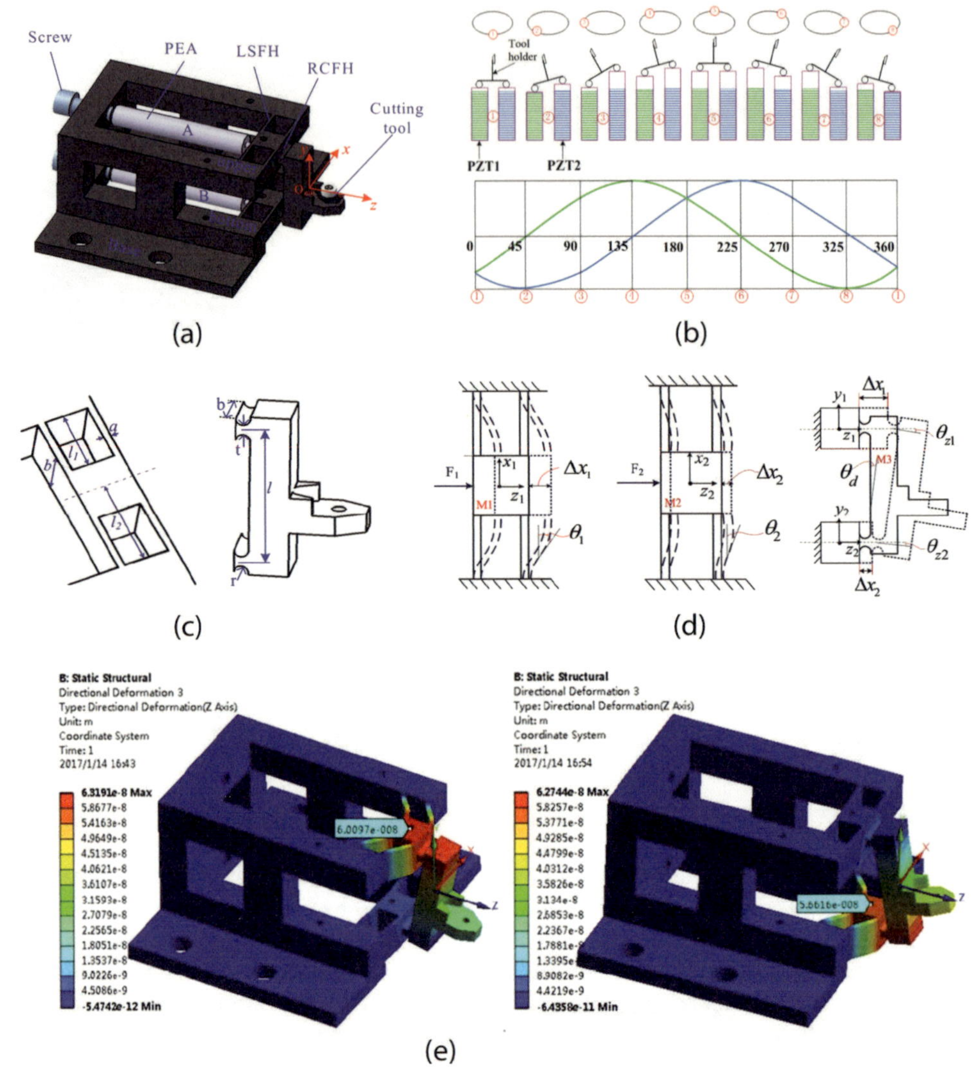

Figure 2.
Illustration of the proposed hybrid flexure hinge EVC actuated by a parallel piezoelectric actuator. (a) Mechanical structure. (b) Working principle. (c) Two types of flexure hinge. (d) Deformation schematic of flexure-based mechanism. (e) FEM simulation [8].

right circular flexure hinges. The double parallel four-bar linkage mechanism is connected in series with the right circular flexure hinges. The upper component and bottom component are connected in parallel with the end-effector beam. The working principle is shown in **Figure 2(b)**. The elliptical locus will be obtained when two electric sinusoidal signals with phase different are applied to the piezoelectric actuators.

The characteristics of the two types of flexure hinge and deformation schematic of flexure-based mechanism are shown in **Figure 2(c)** and **(d)**, respectively. Sixty-five Mn was adopted as the material of main structural components. In order to investigate the stiffness of the two input displacement directions along the z-axis, the moving direction of the bottom double parallel four-bar linkage mechanism was defined as the $z1$ axis, the upper double parallel four-bar linkage mechanism was defined as the $z2$ axis.

The stiffness of the guide mechanism in the $z2$ axial direction can be considered as a statically indeterminate beam. Then the analytical stiffness modeling was carried out. Meanwhile, a constant force of 1 N was imposed on the input end. Simulation result is shown in **Figure 2(e)**. Additionally, the dynamics analysis was also conducted based on the Lagrange equation and FEA method. It can be seen that if the fixed condition was improved, the second natural frequency will become the first natural frequency, which is helpful to achieve a large working bandwidth. The results of stiffness and natural frequency along the working direction both from analytical modeling and FEA were obtained. The comparison results of FEA and analytical modeling are shown in **Table 1**.

According to the analysis above, a prototype was fabricated. Offline tests were carried out to evaluate the performances. The experimental setup is shown in **Figure 3(a)**. As shown in **Figure 3(b)** are the results of the step responses along $z1$ axis and $z2$ axis. The rising time are 1.9 ms and 1.5 ms, and the setting time are respectively 3.43 ms and 3.64 ms. There are no steady errors. The amplitude-frequency responses are shown in **Figure 3(c)**. It can be seen that the first natural frequency is about 1200 Hz. The motion stroke and resolution tests are respectively shown in **Figure 3(d)** and **(e)**, the maximum motion stroke of $z1$ axis is about 15 μm, the maximum motion stroke of $z2$ axis is about 19 μm. The resolution of $z1$ and $z2$ axes are approximately 15 nm and 30 nm. **Figure 3(f)** shows the real experimental tooltip displacement in different phase shifts, which has a good agreement with the simulation results shown in **Figure 3(g)**. Additionally, the developed EVC vibrator has a good tracking performance and very low crosstalk between the motion axes. However, this vibrator can only be used in orthogonal EVC. The mechanical structure needs further optimization to obtain better performance.

3.1.1.2 Improved type

An improved EVC vibrator was developed aiming to solve the problems discussed above. The mechanical structure of compliant mechanism and the prototype are shown in **Figure 4(a)** and **(b)**. This vibrator not only can be used in orthogonal EVC but also can be used in oblique EVC through angle adjustment of the torque gauge [9]. In order to obtain the best performance, three structure parameters were needed to be further optimized considering the compact structure, high motion stroke and working bandwidth.

The optimization problem was stated first. Objective: $\Delta L(t_1, t_2, r) \geq 30\,\mu m$, $f(t_1, t_2, r) > 3000$ Hz. Constraints: $\Delta Lc/\Delta L < 5\%$, $\sigma_{max} < \sigma_s / n$. Within ranges: $t_1 \in [0.5, 1.5]$ mm, $t_2 \in [0.8, 2.4]$ mm, $r \in [0.5, 1.5]$ mm. In this work, the cutting force was taken into consideration during the structure optimization. A response surface methodology was adopted to establish the relationship between the input variables and output parameters. An NSGA-II algorithm was used to perform the optimization process. The main process of the optimization can be concluded as follows:

	Stiffness (N μm^{-1})		Natural Frequency (Hz)	
	$z1$ axis	$z2$ axis	3rd	4th
FEA	17.66	16.64	3613.1	4455.1
Analytical model	16.1	16.1	4640.3	4784.7
Deviation (%)	8.8%	3.2%	28.4%	7.4%

Table 1.
Comparisons results of analytical modeling and FEA method.

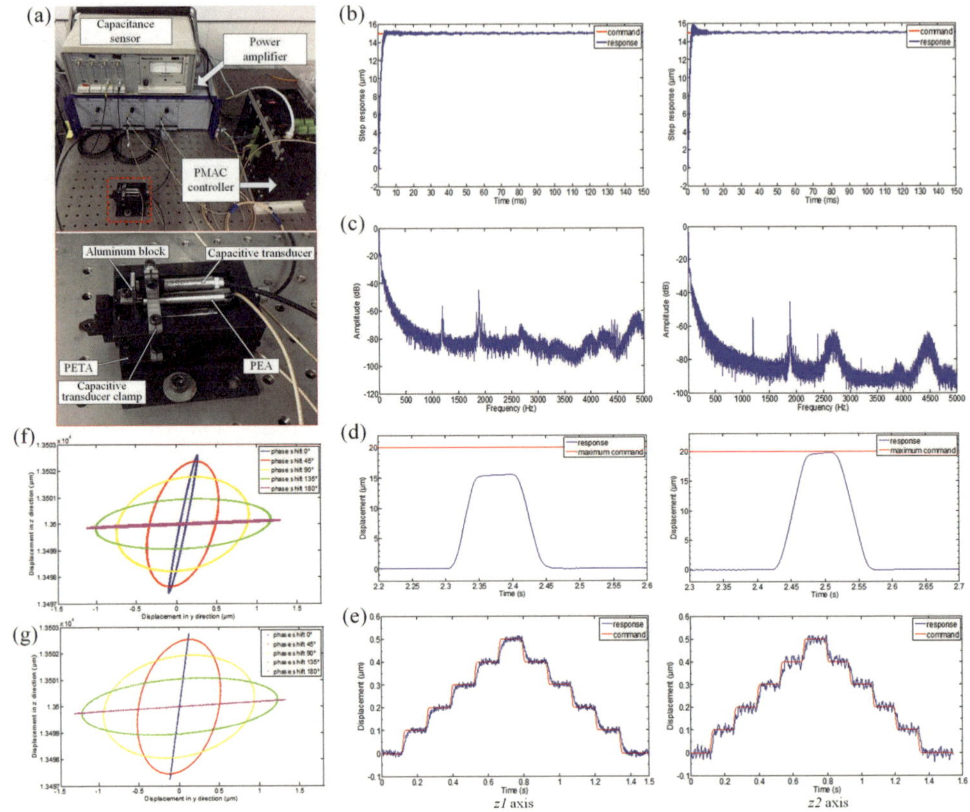

Figure 3.
Experimental setup and testing results. (a) Experimental setup. (b) Step responses. (c) Amplitude-frequency responses. (d) Motion stroke. (e) Resolution tests. (f) Experimental results of output tool tip locus with phase shifts. (g) Simulation results of output tool tip locus with phase shifts [8].

Step 1: Mechanical design is finished first, static and modal analyses are conducted to obtain the response value for the initial design parameters.

Step 2: A response surfaces methodology is adopted to create a predictive model for the design points and the response values. Then, the predicted error should be checked, the other design of experiments methodology or increased experimental design points should be considered when the error is larger than the requirements.

Step 3: MOGA is adopted to deal with the optimization processes via selection, crossover, and mutation. The optimization is converged when the maximum allowable Pareto percentage is realized.

According to the optimization results, a prototype was fabricated with the 7075 T6 aluminum. Offline tests were carried out to evaluate the performance of the optimization vibrator. The results of the step responses along *z1* axis and *z2* axis are shown in **Figure 5(a)**. The rising times are 4.4 ms and 4.2 ms. The setting times are respectively 6.7 ms and 9.06 ms. There are no steady errors and overshoots. The amplitude-frequency responses are shown in **Figure 5(b)**. It can be seen that the first natural frequency is about 1800 Hz. The motion stroke and resolution tests are respectively shown in **Figure 5(c)** and **(d)** by using stair excitation signal to each axis, the maximum motion stroke of *z1* axis is about 37 μm, the maximum motion stroke of *z2* axis is about 31 μm. The resolution of the *z1* axis and *z2* axis are

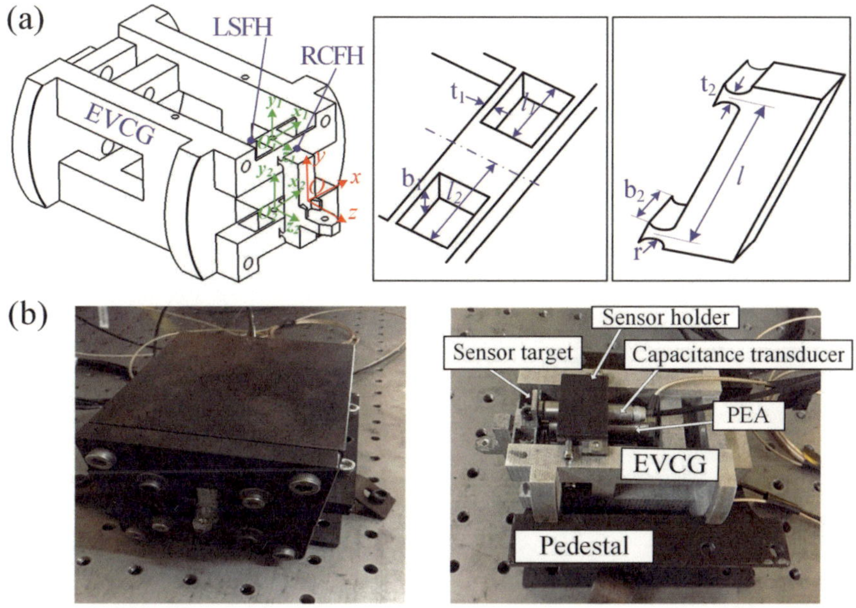

Figure 4.
Illustration of the improved EVC device. (a) The compliant mechanism. (b) The prototype [9].

approximately 9 nm and 10 nm. **Figure 5(e)** shows the motion tracking performance for *z1* axis and parasitic motion for *z2* axis. It can be seen that the maximum tracking error along *z1* axis is 0.7 μm, which is 2.9% of the maximum input displacement. As shown in **Figure 5(f)**, the maximum parasitic motion of *z2* axis is 0.05 μm, which is 0.21% of the maximum input displacement of *z1* axis. For *z2* axis, the maximum tracking error is 0.72 μm, which is 3% of the maximum input displacement. The maximum parasitic motion of *z1* axis is within 0.035 μm, which is about 0.15% of the maximum input displacement. **Figure 5(g)** shows the input signal, the tool vibration locus in 3D space and the projection in *xy* plane. Compared with the former one, this vibrator has higher modulation ability and commonality for lathes with different configurations.

3.1.2 A flexure-based EVC vibrator actuated by vertical piezoelectric actuator configuration

3.1.2.1 Piezoelectric actuators serial drive

In this work, a flexure-based EVC vibrator with vertical piezoelectric actuator configuration was proposed. In order to avoid the installation error, the base part and the hinge part were designed to be an integral flexure hinge structure [10]. Additionally, the integral flexure part was made from 65 Mn steel. As shown in **Figure 6(a)**, the compliant mechanism was driven by two vertical piezoelectric actuators in serial drive mode. The advantage of this serial design is that the crosstalk between two motion axes can be ignored theoretically. However, the working bandwidth decreases with the increased moving inertia.

The testing experiments were carried out to evaluate the performance. As shown in **Figure 6(b)**, the first natural frequencies along Z direction and Y direction are

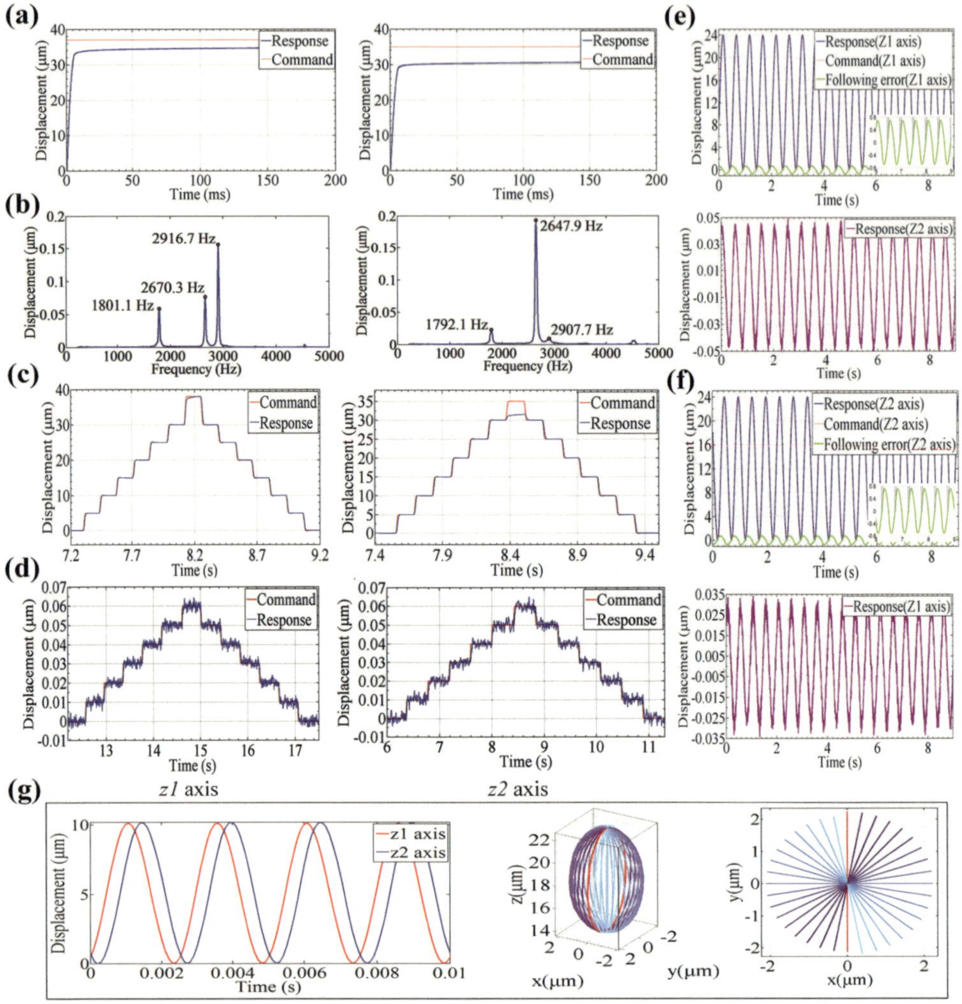

Figure 5.
Offline testing results. (a) Step responses. (b) Amplitude-frequency responses. (c) Motion stroke. (d) Resolution tests. (e) Motion tracking for z1 axis and parasitic motion for z2 axis. (f) Motion tracking for z2 axis and parasitic motion for z1 axis. (g) Input signals and tool vibration locus [9].

1257.74 Hz and 1896.19 Hz, respectively. It can be found from **Figure 6(c)** that the resolution in Z direction and Y direction are all within 8 nm. **Figure 6(d)** and **(e)** show the tracking accuracy and coupling motion in Z direction and Y direction. The maximum following error in Z direction is 0.6 μm, which is about 1.5% of the full testing stroke. The coupling motion in Y direction is 0.06 μm, which is 0.6% of the Y-direction testing stroke. The maximum following error in Y direction is less than 0.2 μm, which is 2% of the testing stroke. The coupling motion in Z-direction can be considered as noise and ignored. **Figure 6(f)** shows the resultant ellipses with different phase shifts and frequencies. It should be noted that the errors in the two directions ascend with the frequency rising. The error in Z-direction is smaller than that in Y-direction. **Figure 6(g)** shows the resultant tool paths with different phase shifts under-cutting speed of 150 μm/s. **Figure 6(h)** shows the resultant tool paths

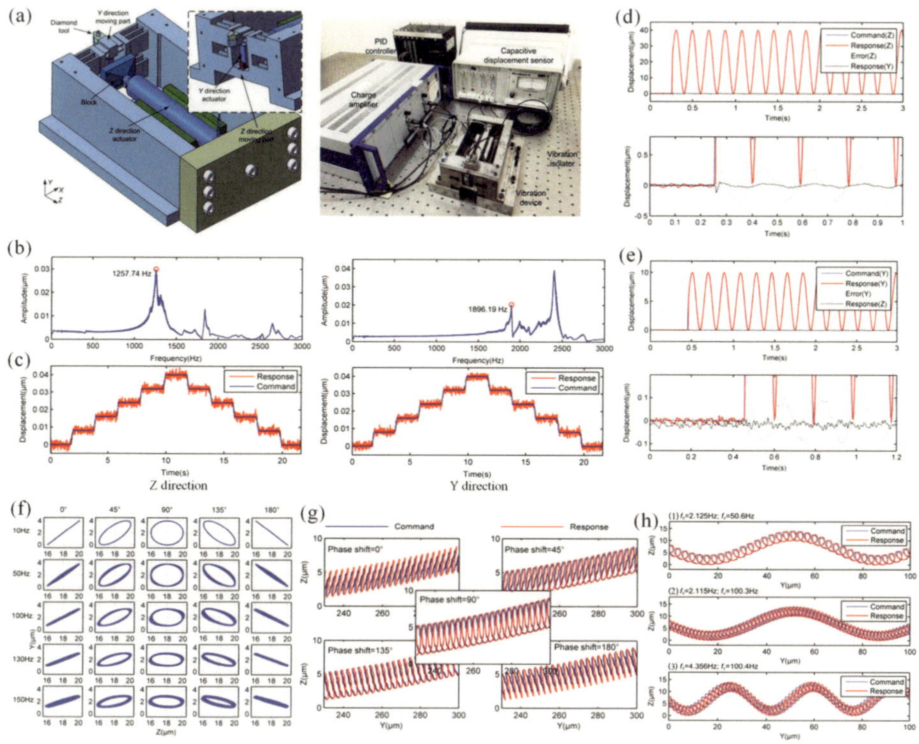

Figure 6.
Illustration of mechanical structure, experimental setup and testing results. (a) Mechanical structure and experimental setup. (b) Amplitude-frequency responses. (c) Resolution tests. (d) Motion tracking for Z-axis and parasitic motion for Y-axis. (e) Motion tracking for Y-axis and parasitic motion for Z-axis. (f) Resultant ellipses with different phase shifts and frequencies. (g) Resultant tool paths with different phase shifts. (h) Resultant tool paths at different frequencies [10].

with different frequencies under cutting speed of 150 μm/s and 90° phase shift. These two figures demonstrate the double-frequency EVC tracking ability.

3.1.2.2 Piezoelectric actuators parallel drive

Compared with the piezoelectric actuators' serial drive configuration, the parallel drive configuration of piezoelectric actuators is more easy to obtain a higher working bandwidth. However, the cross-axis coupling is a problem that need to be solved.

In general, the mechanical structure decoupling design is usually adopted. In this work, a modified bridge-type amplification mechanism was utilized to meet the requirements among the stroke, output stiffness, resonant frequency, and the drive current [11]. Spring steel was adopted as the material. As shown in **Figure 7(a)**, two perpendicular leaf-spring flexure hinges were applied to decouple the 2D motion. **Figure 7(b)** shows the FEA results of modal analysis. The natural frequency is 6452 Hz and the corresponding mode of vibration is in cutting direction. The mode of vibration in thrust direction corresponds to the natural frequency of 7432 Hz.

In order to evaluate the performance of the prototype. The experimental test was conducted and the experimental setup is shown in **Figure 7(c)**. The sweep experiments were conducted by setting the input voltage amplitudes to 5 V and

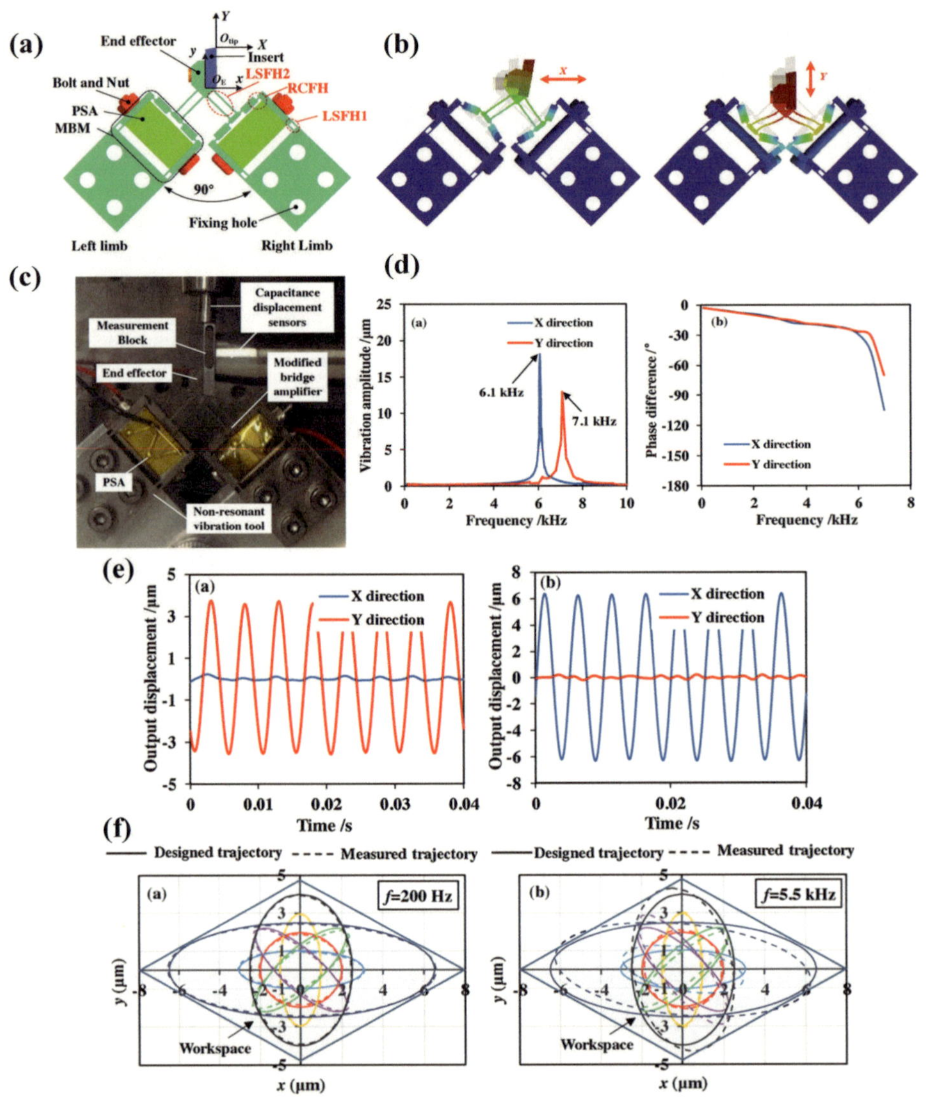

Figure 7.
Illustration of mechanical structure, experimental setup, and testing results. (a) Mechanical structure. (b) Modal analysis results. (c) Experimental setup. (d) Frequency response. (e) axis coupling results. (f) Comparison between designed and measured cutting tool locus [11].

the frequency from 0 to 10 kHz linearly. The amplitude-frequency responses and phase-frequency responses are shown in **Figure 7(d)**. The natural frequencies in X and Y directions (i.e. cutting direction and thrust direction) are 6100 Hz and 7100 Hz, respectively, which have good agreement with the results of FEA simulation. Coupling tests of the two motion axes were performed by using sinusoidal signals with frequency of 200 Hz and amplitude of 120 V. The results are shown in **Figure 7(e)** that the coupling ratios between two motion axes are within 5%. The comparison between designed and measured cutting tool locus are shown in **Figure 7(f)** for both low-frequency (200 Hz) and high-frequency (5.5 kHz). The measured results indicate that the agreement between designed and measured ones are good for low frequency.

However, there is a discrepancy between designed and measured ones under higher frequency caused by linearly increasing phase shift which can be eliminated by frequency compensation.

In fact, the angle between two decoupling flexure hinges can influence the decouple performance. A new type of 2 degrees of freedom piezo-actuated pseudo-decouple compliant mechanism was developed which considering the influence of the decoupling angle [12]. As shown in **Figures 8(a)** and **8(c)**, the two perpendicular piezoelectric actuators are configured in parallel. In this work, the influence of decoupling angle Θ on tracking accuracy of elliptical locus was studied by static FEA method. First, a dimensionless aspect ratio λ of the major semi-axis a to the minor semi-axis b of the ellipse locus was introduced. The influence of decoupling angle on the elliptical parameters and relative ratio is shown in **Figure 8(b)**. It can be seen that the developed 2D pseudo-decouple compliant mechanism generated an approximately perfect ellipse locus when the decoupling angle was set to be 102.5°. Then a 3D model of EVC vibrator was established as shown in **Figure 8(c)**. The prototype was manufactured and the experimental tests were carried out to evaluate the performance. Experimental setup is shown in **Figure 8(d)**. **Figure 8(e)** shows the experimental results of kinematic performance. From the resultant tool locus, it can

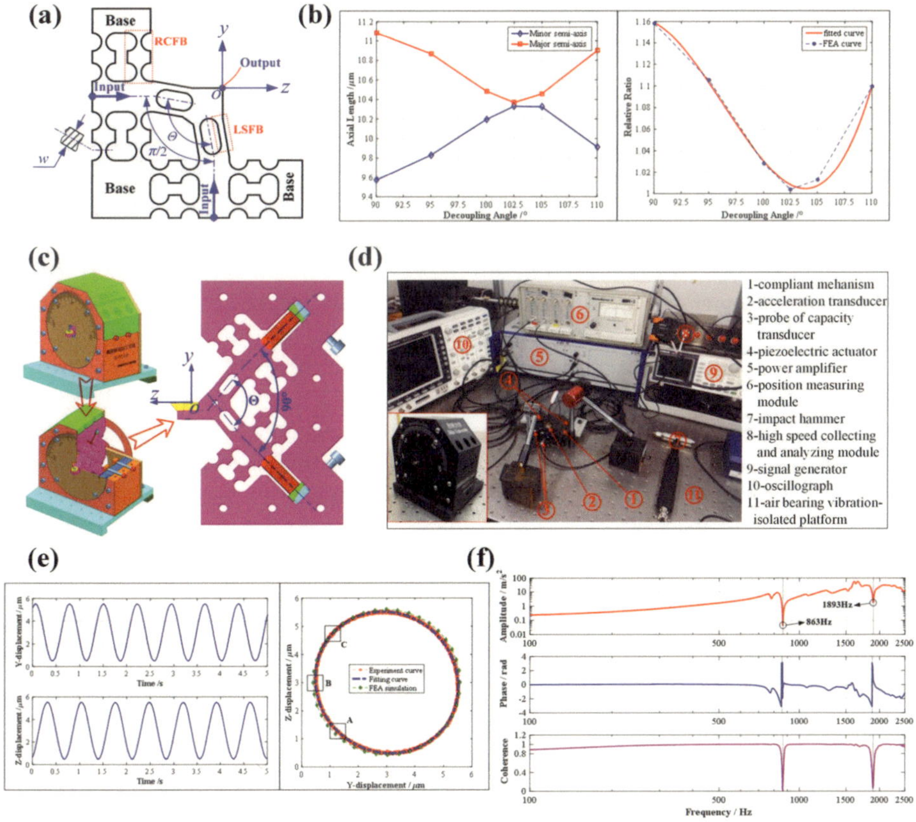

Figure 8.
Illustration of mechanical structure, experimental setup, and testing results. (a) Mechanical structure of compliant mechanism. (b) Influence of decoupling angle on the elliptical parameters and relative ratio. (c) 3D model of the EVC vibrator. (d) Experimental setup. (e) Experimental results of kinematic performance. (f) Experimental results of dynamic performance [12].

be seen that the experimental results have a good agreement with the fitted curve and FEA results. **Figure 8(f)** shows the results of dynamic performance by using hammering method. The first two natural frequencies of the pseudo-decouple compliant mechanism are 863 Hz and 1893 Hz, respectively.

3.2 3D type EVC vibrator

3D type EVC vibrator has more degrees of freedom compared with 2D type EVC vibrator. Thus, it has more flexibility and modulation ability to fabricate the structured surface than 2D type EVC vibrator. However, the structure was more complicated and it is difficult to control the structure size, which may influence the potential applications. In recent 10 years, the flexure-based 3D type EVC vibrator developed rapidly. In general, there are three commonly used piezoelectric actuator configurations, i.e. three perpendicular piezoelectric actuators configuration, two parallel and one perpendicular piezoelectric actuators configuration, and four parallel piezoelectric actuators configuration. Here are some typical examples following below.

3.2.1 A flexure-based EVC vibrator actuated by three perpendicular piezoelectric actuators

For 3D type EVC vibrator, three perpendicular piezoelectric actuators configuration are the most commonly used method. A piezoelectric actuated monolithic compliant spatial vibrator with decoupled translational vibration was developed to construct the rotary spatial vibration-assisted diamond cutting system [13]. The 3D model of the rotary spatial vibration-assisted diamond cutting system and the mechanical structure of compliant spatial vibrator are respectively shown in **Figures 9(a)** and **9(b)**. The rotary spatial vibration-assisted diamond cutting system consists of a tool holder, compliant spatial vibrator, piezoelectric actuator, compliant spatial vibrator holder, connection shaft, and fixture. The fixture is used for attaching the whole mechanism onto the spindle of the machine tool through a vacuum chuck. The compliant spatial vibrator and its holder are made of aluminum alloy to reduce the mass of the whole mechanism. The steel was adopted for connecting shaft manufacturing to increase connection stiffness. In this work, a complete compliance modeling was established based on the matrix-based compliance modeling method for compliant spatial vibrator. The dynamic model was established based on Lagrangian principle. FEA simulation was also used to study the static and dynamic characteristics. To evaluate the stroke and parasitic motions, a maximum sinusoid driving voltage ($u = [50 + 50\sin(2\pi t)]$ V) was separately applied to each piezoelectric actuator. The displacement results of compliant spatial vibrator in driving direction and the parasitic motion in other motion directions are shown in **Figure 9(c)**. It can be seen that the practical stroke can reach 11.067 μm, 10.100 μm, and 12.254 μm along the x, y, and z axes directions, respectively. The parasitic motions along y and z axes directions are about ±1.39% and 1.34% with respect to the motion along x-axis direction, respectively. Similarly, the parasitic motions along the x and z axes directions are about ±1.11% and 0.46% with respect to the motion along y-axis direction, respectively. The parasitic motions along the x and y axes directions are about ±1.21% and 0.31% with respect to the motion along z-axis direction, respectively. In addition, the dynamic performance tests were carried out by swept excitation method. Signals with an amplitude of 1.5 V with varying frequency were applied to each piezoelectric actuator, separately. The results of

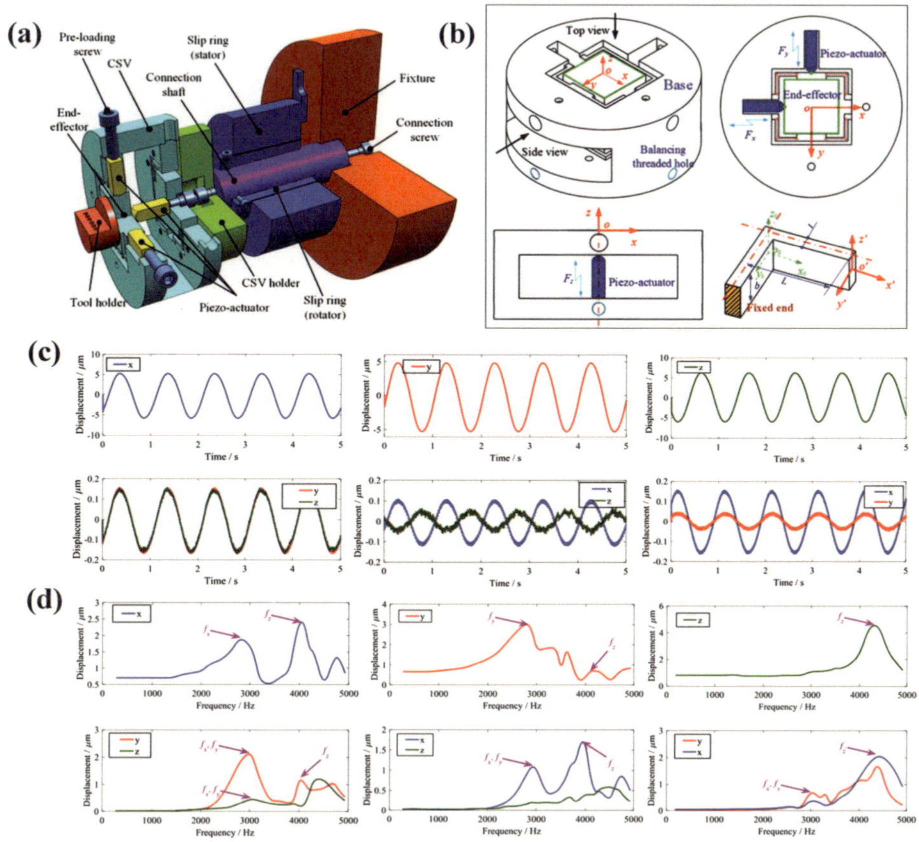

Figure 9.
Illustration of 3D model, mechanical structure and testing results. (a) 3D model of the rotary spatial vibration-assisted diamond cutting system. (b) Mechanical structure of the compliant spatial vibrator. (c) Displacement results of compliant spatial vibrator in driving direction and the parasitic motion in other motion directions. (d) Experimental results of dynamic performance [13].

dynamic performance are shown in **Figure 9(d)**. The first natural frequency along x and y axes directions are approximately the same, which are about $f_x = f_y = 2.8$ kHz. While the natural frequency along the z-axis direction is about 4.3 kHz. This vibrator provides a new method for rotary vibration machining.

Different from the above 3D type EVC vibrator, a piezo-actuated tri-axial compliant mechanism was developed to modulate the workpiece for nano cutting [14]. The 3D model of the compliant mechanism is shown in **Figure 10(a)**. Three identical compliant chains were adopted to construct the vibrator. For one compliant chain, the double parallelogram mechanism with eight right circular flexure hinges and a spatial transition mechanism with four sets of parallelogram limbs were adopted. Each limb has two bi-axial right circular flexure hinges. In this work, multi-objective optimal design of the tri-axial compliant mechanism was carried out based on the stiffness modeling, kinematic modeling and dynamic modeling. A Pareto-based multi-objective differential evolution algorithm was utilized to find the global optimal solution.

Then a prototype was manufactured and the performance tests were conducted through the experimental setup which is shown in **Figure 10(b)**. A harmonic signal with frequency of 1 Hz and amplitude voltage of 5 V was applied to each actuator

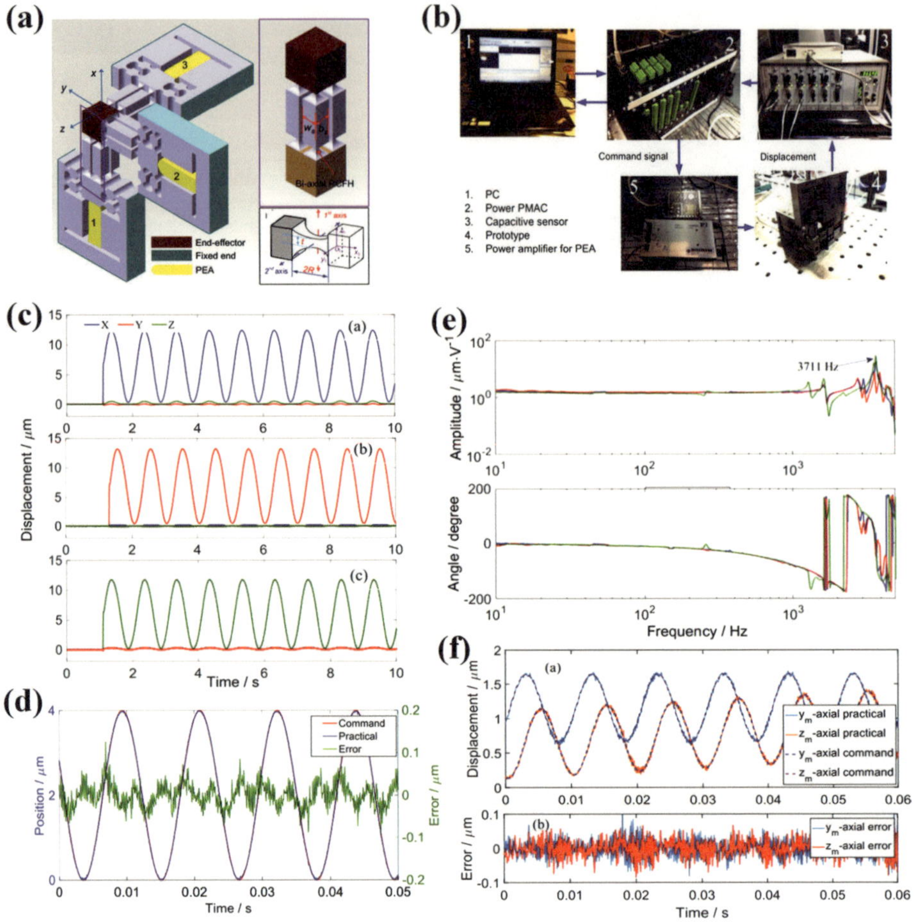

Figure 10.
Illustration of 3D model, experimental setup and testing results. (a) 3D model of the compliant mechanism. (b) Testing experimental setup. (c) Results of harmonic response and the corresponding parasitic motions. (d) Harmonic tracking performance for x-axis. (e) Results of the frequency responses. (f) Harmonic tracking performance for y and z axes during cutting [14].

to assess the motion stroke and the parasitic motion. The results are shown in **Figure 10(c)**, the strokes along x, y and z axial directions are respectively 12.37 μm, 13.14 μm and 11.72 μm. The maximum parasitic motion (about 3.8%) was generated along z axial direction when driving the piezoelectric actuator along x axial direction. When actuated the piezoelectric actuators along with y and z axial directions, the parasitic motion along the other two directions are almost the same, which are about 1.37% and 2.51% of the corresponding actuation motions. In addition, a typical proportion-integration-differentiation controller is employed for the feedback control. The influences of system noises were eliminated by a low pass filter, and a velocity feedforward compensator was utilized to enhance the response speed. In order to assess the tracking performance, a harmonic signal with frequency of 90 Hz and amplitude of 2 μm was adopted as the desired motion. Only the tracking performance along x axial direction was performed to avoid repetition. The tracking results in **Figure 10(d)** shows a tracking error of around ±70 nm, which is about ±1.75% of the motion span. **Figure 10(e)** shows the dynamic performances. The natural frequency for each direction is around 3.7 kHz due to the identical chain. What's more, the tracking performance along *y* and *z* axial directions were also tested during cutting, results of which are shown in **Figure 10(f)**. It should be noted that the noise during cutting is slightly larger than that in offline testing.

3.2.2 A flexure-based EVC vibrator actuated by two parallel and one perpendicular piezoelectric actuators

In this work, a flexure-based EVC vibrator actuated by two parallel and one perpendicular piezoelectric actuator was proposed [15]. The 3D model of the developed

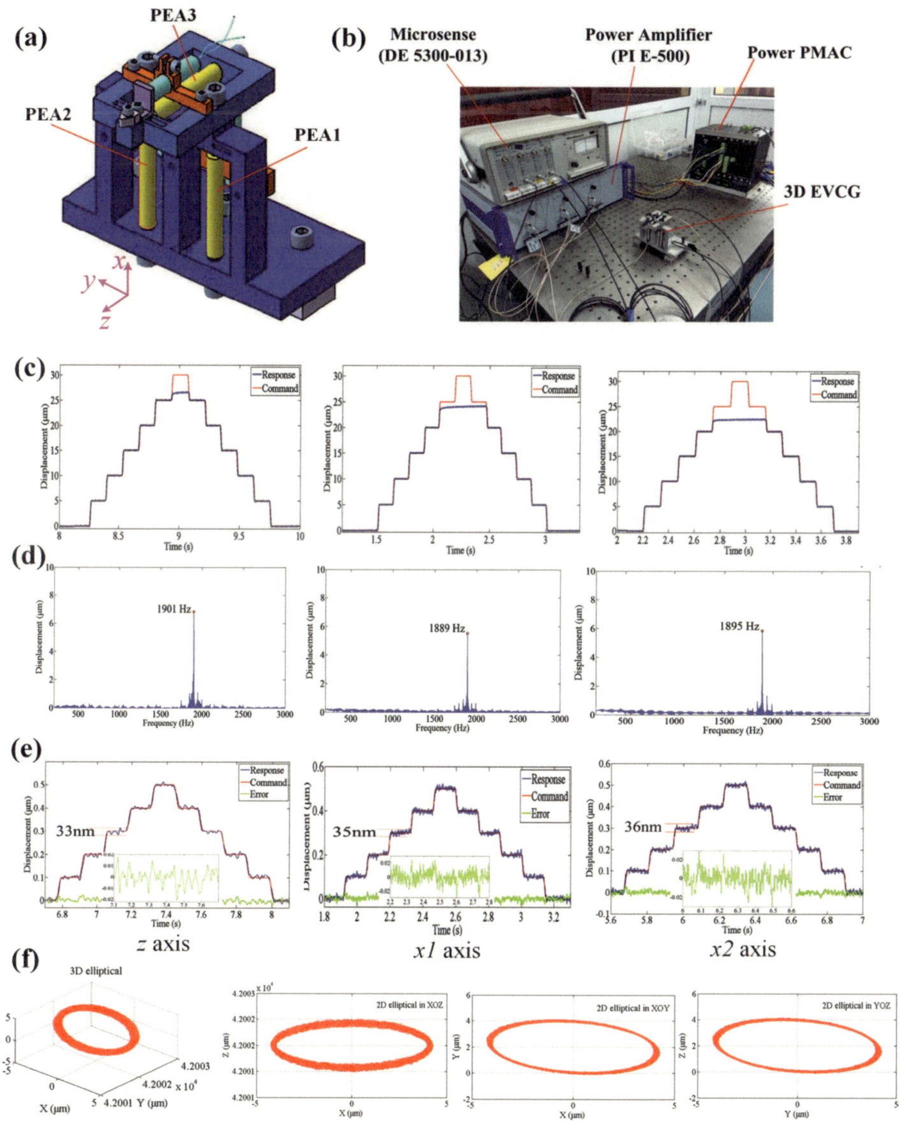

Figure 11.
Illustration of 3D model, experimental setup and testing results. (a) 3D model of the developed EVC vibrator. (b) Testing experimental setup. (c) Results of motion stroke. (d) Results of the frequency responses. (e) Results of the resolution test. (f) the synthesized cutting tool locus and the projections [15].

EVC vibrator is shown in **Figure 11(a)**. The EVC vibrator mainly consists of three PEAs (piezoelectric actuators) and capacity probes, two compliant mechanisms. Two compliant mechanisms were independently manufactured considering the complexity of machining. Then the compliant mechanisms were assembled by screws connection. For easy analysis, three axes were defined as follows: axis along PEA1 motion direction was defined as *x1*, axis along PEA2 motion direction was defined as *x2*,

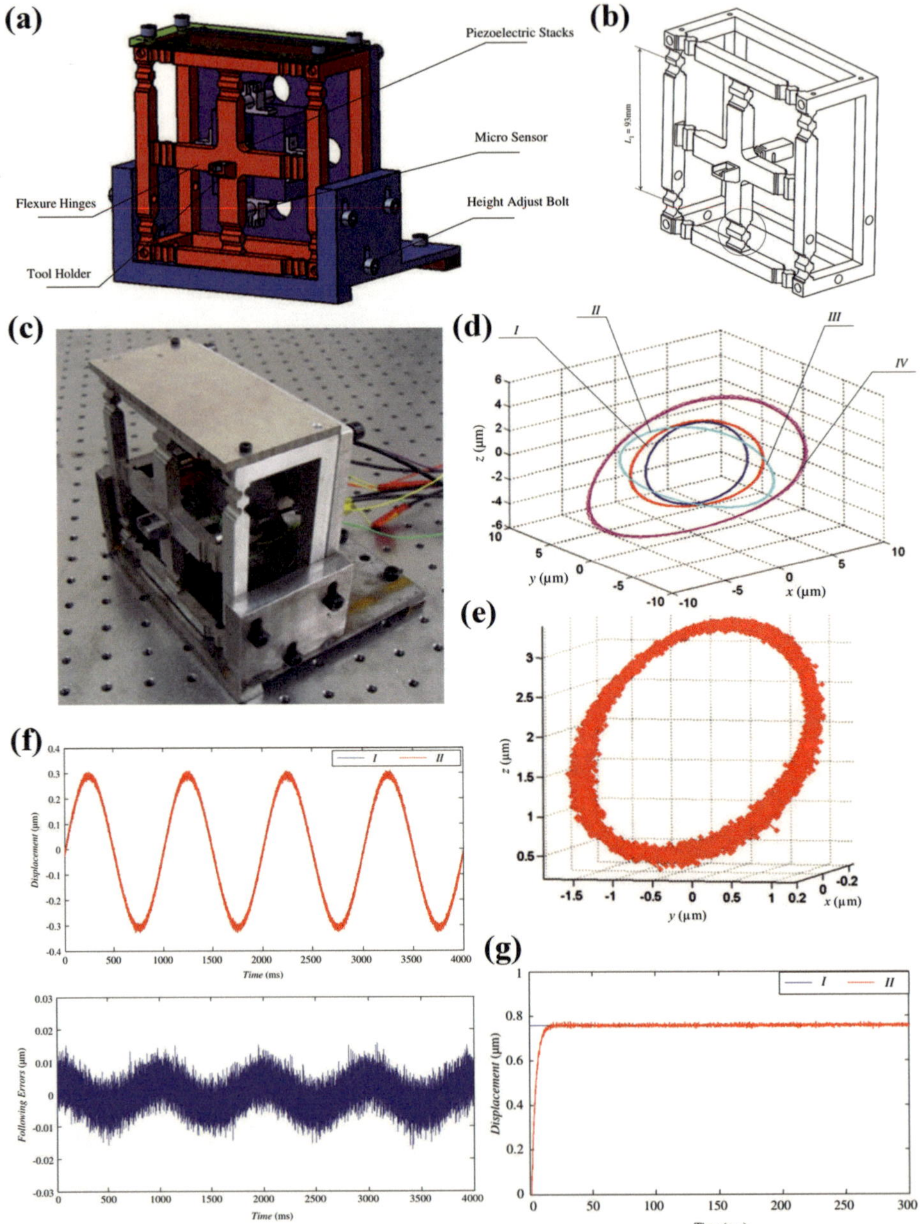

Figure 12.
Illustration of 3D model, prototype and testing results. (a) 3D model of the developed EVC vibrator. (b) Mechanical structure of compliant mechanism. (c) Fabricated prototype. (d) Simulated tool locus. (e) Actual tool locus. (f) Tracking performance. (g) Step response [16].

and axis along PEA3 motion direction was defined as z. Additionally, the kinematic modeling was established. The offline tests were carried out to investigate the performance based on the experimental setup shown in **Figure 11(b)**. The motion stroke performances along three axes are shown in **Figure 11(c)**. It can be seen that the maximum displacement along z-axis can reach up to 26 μm. The maximum displacements along $x1$ and $x2$ axes are not the same which are 22 μm and 24 μm, respectively, due to the manufacturing error. A swept excitation method with frequency from 100 Hz to 3000 Hz was performed to assess the dynamic characteristics. **Figure 11(d)** shows that the natural frequency along z, $x1$ and $x2$ axes are about 1901 Hz, 1889 Hz and 1895 Hz, respectively, which is enough for ultra-precision machining. The resolution performance results are shown in **Figure 11(e)** through stair excitation tests. The resolution of motion axes along z, $x1$ and $x2$ axes are about 33 nm, 35 nm and 36 nm, respectively. Meanwhile, the synthesized cutting tool locus and the projections are shown in **Figure 11(f)** which validate the correctness of the kinematic model and feasibility of the developed EVC vibrator.

3.2.3 A flexure-based EVC vibrator actuated by four parallel piezoelectric actuators

In this work, a flexure-based EVC vibrator actuated by four parallel piezoelectric actuators was proposed [16]. The mechanical structure of the developed EVC vibrator is shown in **Figure 12(a)**. It consists of one fixture base, one compliant mechanism, four piezoelectric actuators and capacity sensors. The compliant mechanism is shown in **Figure 12(b)**, which was fabricated by wire EDM. The prototype of developed EVC vibrator was fabricated and assembled as shown in **Figure 12(c)**. In order to investigate the kinematic characteristics, the kinematic model was established and the simulation results are shown in **Figure 12(d)**, of which I represents reference ellipse, II represents ellipse obtained by changing the acting locations of the four piezoelectric actuators, III represents ellipse obtained by changing the phase shifts of actuated signals, IV represents ellipse obtained by changing the amplitudes of actuated signals. **Figure 12(e)** shows the measured tool locus which is an ellipse in 3D space. The measured tool locus are in accordance with the simulated locus shown in **Figure 12(d)**. **Figure 12(f)** shows the tracking performance in x-direction. A sine wave with amplitude of 6 μm and frequency of 1 Hz was adopted as the command signal. The maximum following error is about 12 nm, which is less than 4% of the full span. The result of step response along x-direction is shown in **Figure 12(g)**. There are almost no overshoot and steady errors by utilizing a typical PID controller.

4. Conclusions

The structured surfaces with various micro/nano characteristics may have different advanced functions in various fields. However, the machining of micro/nano structured surfaces is becoming a challenge for traditional cutting methods on the difficult-to-cut materials. This chapter focus on the EVC technique which is a promising way for structured surface machining. The advantages of EVC were described in detail. Meanwhile, the methods for realization of EVC were also introduced in detail. As is known, the flexure-based non-resonant type EVC vibrator has more flexibility and modulation ability compared with the resonant type EVC vibrator and it has been widely concentrated in recent years. Therefore, this chapter mainly reviews the recent developments and achievements on flexure-based EVC vibrators actuated by piezoelectric actuators. According to the locus characteristics of cutting tool and its position and posture, the EVC was categorized into three different types. And this chapter reviews some typical examples according to the different types of EVC, including the mechanical structure, working principle of piezoelectric actuator, the static modeling, kinematic and dynamic modeling, structure optimization and testing. Although the flexure-based EVC vibrator developed rapidly, there are still some important problems that severely restricted the performance of EVC vibrator, which need to be further investigated in the future, such as the conflict between high working bandwidth and large working stroke, topological optimization of mechanical structures, fatigue life problems under long working hours, and design of high-performance controller.

Acknowledgements

This work was financially supported by the National Natural Science Foundation of China (Grant No. 51905322) and the Postdoctoral Science Foundation of China (Grant No. 2021 T140420).

Conflict of interest

The authors declared no potential conflicts of interest with respect to the research, authorship, and/or publication of this article.

Author details

Jinguo Han[1*], Mingming Lu[2] and Jieqiong Lin[2]

1 School of mechanical engineering, Shandong University of Technology, Zibo, Shandong, China

2 Key Laboratory of Micro-Nano and Ultra-Precision Manufacturing of Jilin Province, School of Mechatronic Engineering, Changchun University of Technology, Changchun, Jilin, China

*Address all correspondence to: hankeyee@163.com

References

[1] Zheng L, Chen W, Huo D. Review of vibration devices for vibration-assisted machining. The International Journal of Advanced Manufacturing Technology. 2020;**108**:1631-1651. DOI: 10.1007/s00170-020-05483-8

[2] Zhu Z, To S, Xiao G, et al. Rotary spatial vibration-assisted diamond cutting of brittle materials. Precision Engineering. 2016;**44**:211-219. DOI: 10.1016/j.precisioneng.2015.12.007

[3] Zhang J, Zhang J, Liu C, et al. Machinability of single crystal calcium fluoride by applying elliptical vibration diamond cutting. Precision Engineering. 2020;**66**:306-314. DOI: 10.1016/j.precisioneng.2020.06.008

[4] Zhang J, Han L, Zhang J, et al. Brittle-to-ductile transition in elliptical vibration-assisted diamond cutting of reaction-bonded silicon carbide. Journal of Manufacturing Processes. 2019;**45**:670-681. DOI: 10.1016/j.jmapro.2019.08.005

[5] Shamoto E, Moriwaki T. Study on elliptical vibration cutting. CIRP Annals. 1994;**43**(1):35-38. DOI: 10.1016/S0007-8506(07)62158-1

[6] Cerniway M A. Elliptical diamond milling: kinematics, force and tool wear. Master Thesis. North Carolina State University. 2002

[7] Negishi N. Elliptical vibration assisted machining with single crystal diamond tools. Master Thesis. North Carolina State University. 2003

[8] Lin J, Han J, Lu M, et al. Design, analysis and testing of a new piezoelectric tool actuator for elliptical vibration turning. Smart Materials and Structures. 2017;**26**(8):085008. DOI: 10.1088/1361-665X/aa71f0

[9] Han J, Lin J, Li Z, et al. Design and computational optimization of elliptical vibration-assisted cutting system with a novel flexure structure. IEEE Transactions on Industrial Electronics. 2018;**66**(2):1151-1161. DOI: 10.1109/TIE.2018.2835425

[10] Zhou X, Zuo C, Liu Q, et al. Development of a double-frequency elliptical vibration cutting apparatus for freeform surface diamond machining. The International Journal of Advanced Manufacturing Technology. 2016;**87**(5):2099-2111. DOI: 10.1007/s00170-016-8596-2

[11] Wang J, Du H, Gao S, et al. An ultrafast 2-D non-resonant cutting tool for texturing micro-structured surfaces. Journal of Manufacturing Processes. 2019;**48**:86-97. DOI: 10.1016/j.jmapro.2019.10.023

[12] Wang R, Zhou X, Meng G. Development of a new type of 2-DOF piezo-actuated Pseudo-decoupled compliant mechanism for elliptical vibration machining. Micromachines. 2019;**10**(2):122. DOI: 10.3390/mi10020122

[13] Zhu Z, To S, Ehmann KF, et al. Design, analysis, and realization of a novel piezoelectrically actuated rotary spatial vibration system for micro−/nanomachining. IEEE/ASME Transactions on Mechatronics. 2017;**22**(3):1227-1237. DOI: 10.1109/TMECH.2017.2682983

[14] Zhu Z, To S, Zhu WL, et al. Optimum design of a piezo-actuated triaxial compliant mechanism for nanocutting.

IEEE Transactions on Industrial Electronics. 2017;**65**(8):6362-6371. DOI: 10.1109/TIE.2017.2787592

[15] Lin J, Han J, Lu M, et al. Design and performance testing of a novel three-dimensional elliptical vibration turning device. Micromachines. 2017;**8**(10):305. DOI: 10.3390/mi8100305

[16] Lin J, Lu M, Zhou X. Development of a non-resonant 3D elliptical vibration cutting apparatus for diamond turning. Experimental Techniques. 2016;**40**(1):173-183. DOI: 10.1007/s40799-016-0021-0

Chapter 5

Influence of Piezoelectric Actuator Properties on Design of Micropump Driving Modules

Matej Možek, Borut Pečar, Drago Resnik and Danilo Vrtačnik

Abstract

The chapter will briefly present three distinctive concepts of the micropump actuator driving module, each with its waveform specifics and their impact on particular micropump performance (pumping media, flow rate and backpressure). First presented concept is based on two mutually-exclusive boost switched-mode power supply modules. Characterization of this module identified output voltage asymmetry to be the limiting factor of micropump performance. To assure driving symmetry, an alternative driving module, based on independent high-voltage stages and optocouplers, was implemented. This design is capable of driving a piezoelectric micropump with a rectangular waveform of programmable frequency, positive and negative amplitudes, slew rates and dead time. While this design provides maximum flow and backpressure characteristics, it does not offer minimal current consumption and long-term operation. To overcome this difficulty, our current design is based on an embedded arbitrary waveform generator, which offers an efficient trade-off between high pumping performance and low current consumption.

Keywords: piezoelectric micropumps, mutually exclusive switched-mode power supply modules, optocoupler, arbitrary waveform generator, low current consumption

1. Introduction

Modern microfluidic applications are moving toward miniaturization and lowering current consumption demands. Consequently, digital system voltages are lowered to a level of 3 V and below, which are most commonly used today. On the other hand, piezoelectric micropumps need a high voltage signal in the range of few hundred volts for normal operation. Driving signal parameters are depending on PZT material (Lead zirconate titanate), construction and required performance of a specific micropump. Therefore, a low-power portable piezoelectric micropump driver presents an interesting challenge in electronics design.

In addition to being capable of high voltage waveform synthesis, the piezoelectric micropump driver should allow for the adaptations of signal waveform shape, amplitude and frequency. These parameters need to be optimized for a type of driven micropump in order to maximize the microfluidic system performance.

IntechOpen

Commercially available drivers are often either physically large and therefore non-portable [1] or are dedicated to driving a particular type of piezoelectric micropump [2]. Furthermore, available drivers offer only limited signal flexibility. In our previous work, a 3-channel high voltage AB class linear amplifier was developed [3]. This module offered very good micropump driving signals up to 10 kHz, but it was not size optimized. A miniaturized, high-voltage micropump driver was also implemented using a piezo haptic driver DRV2667 [4]. This implementation featured fully-programmable signal shape, frequency and amplitude, but was limited by signal driving amplitude to 200 V_{pp} and frequency range up to 1 kHz. Current consumption was 134 mA at such high excitation voltages.

The above-listed limitations of referenced drivers encouraged the development of a simpler, cost-effective micropump driver electronic module, which would be limited to a rectangular driving signal, but would offer excitation with higher voltages, whilst maintaining the low-power aspect (i.e. current consumption in the order of tenths of mA). One option was to design the driver with a separate high-voltage power supply, which provides a driving signal using an H-bridge, but due to complexity of such a circuit, this never represented a cost-effective solution. H-bridge topology does not ground one of the micropump actuator terminals.

Low-cost aspect of aforementioned implementations was challenged by a transformerless design, proposed and patented by Fraunhofer IZM [5]. Their brilliantly simple idea features two switched-mode power supply (SMPS) boost converters, which operate in mutual exclusion. Each boost converter forms voltage of either positive/negative polarity, but both SMPS converters incorporate same piezoelectric micropump as its output capacitor. The need for a dedicated output capacitor results in a miniature, digitally controlled version of a piezoelectric micropump driving module. Such interchanging SMPS module design synthesizes rectangular shape of driving signal with resistor-capacitor (RC) charging and discharging transitions through the piezoelectric micropump. Though the resulting edge transitions of a rectangular micropump driving signal are not ideal, the performance of such a circuit can be considered adequate for certain cost- and size-sensitive applications. The primary objective of this chapter is to present the development of three distinctive micropump driving module designs with their impact on piezoelectric micropump electrical and fluidic characteristics.

2. Micropump module, based on mutually exclusive SMPS

The high-voltage section of the analyzed circuit is designed around two boost SMPS power supplies, depicted in **Figure 1**. Each SMPS module provides corresponding positive and negative micropump excitation voltage: Positive voltage SMPS circuit comprises a switching transistor Q_3, a diode D_2, while the negative voltage SMPS comprises the switching transistor Q_4, diode D_1. If the switching operation between SMPS modules is alternated repeatedly, a rectangular shape of driving signal with RC charging and discharging transitions is synthesized. Micropump PZT actuator is represented by C_1 and acts as a shared SMPS output capacitor in both circuits. Boost inductance L_1 is also shared. While the driver is supplying negative voltage to the micropump C_1, the transistor Q_3 is fully open (its base is tied via R_3 to V_{CC}) and the Q_4 with diode D_1 performs the PWM (pulse width modulation) switching. For positive voltage, the roles of Q_3 and Q_4 are reversed: Q_4 is fully open (its base tied via R_4 to GND) and the Q_3 with diode D_1 performs the PWM switching. Transistors Q_1 and Q_2 in common-base orientation provide isolation between the two SMPS stages and prevent micropump capacitor C_1 discharge through the inactive (i.e. fully open) transistor. During positive voltage switching, the Q_4 is fully open - its collector-emitter voltage is

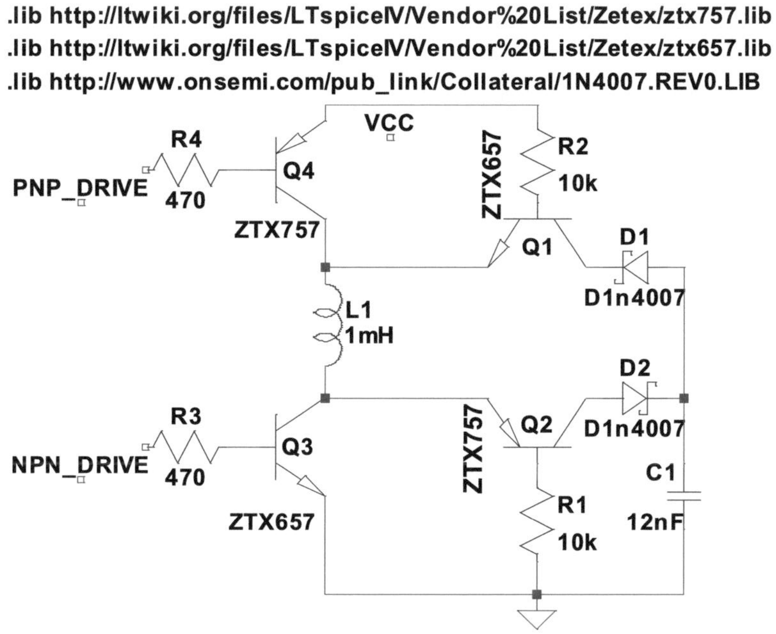

Figure 1.
High-voltage part of the mutually exclusive SMPS driver design.

reduced to its saturation value, hence the Q_1 isolation transistor is closed, thus effectively preventing any reverse current flow through diode D_1. Similar analysis can be applied for transistors Q_2, Q_3 and the diode D_2 during negative voltage switching cycle.

In order to synthesize a rectangular shape of driving signal with resistor-capacitor charging and discharging transitions, both positive and negative SMPS boost circuits have to be switched in the abovementioned sequence by providing the PNP_DRIVE and NPN_DRIVE signals.

To achieve this, an initial, simplified version of the micropump driving circuit was designed to assess the optimal PWM switching frequency range and establish the need for a PWM drive. High voltage part driving circuit, depicted in **Figure 2**, was based around a TTL 74HC257 4-channel 2/1 multiplexer for 5 V power supply version or its CMOS counterpart (40257) for 12 V version. Each 2/1 multiplexer comprises *a* and *b* inputs and an output *Y*. All multiplexers share the same selection input *s*. In **Figure 2**, only lower two multiplexers, indexed *0* and *1*, are used. High-frequency switching is provided by 2 AC voltage sources V_2 and V_4. Micropump excitation frequency signal is provided by V_3, which is connected to the multiplexer selection input. Toggling speed of the multiplexer selection input thus corresponds to mutual exclusion of SMPS power supplies, which directly translate to micropump excitation frequency.

2.1 Circuit simulations

Above described circuit was initially simulated using LTSpice IV [6]. In presented case micropump was modeled using previously measured capacitance C_0 of 12 nF [7]. Synthesized rectangular signal frequency, which represents micropump excitation, was set at 100 Hz with 50% symmetry. High-frequency switching of transistors Q_3 and Q_4 was swept from 10 kHz to 33 kHz, whilst recording resulting micropump excitation voltage. Resulting recorded micropump voltage at 100 Hz is depicted in **Figure 3**.

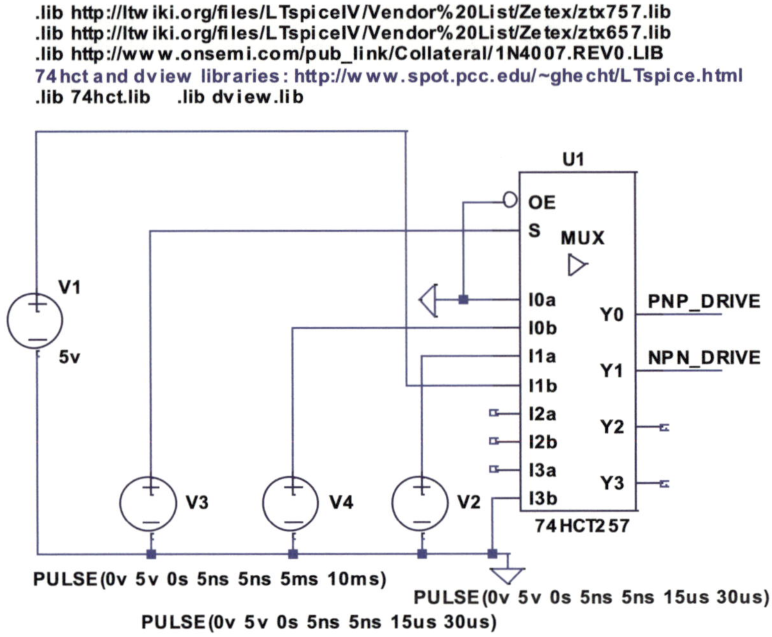

Figure 2.
Excitation part of the mutually exclusive SMPS driver design.

Initial simulations have shown, that the output voltage, depicted in **Figure 3**, features a pronounced, unwanted asymmetry of positive and negative half-cycle amplitudes in order of 50 V. This consequentially causes a DC offset voltage, which deteriorates the piezo actuator deflection over time or even prevents proper micro-pump performance.

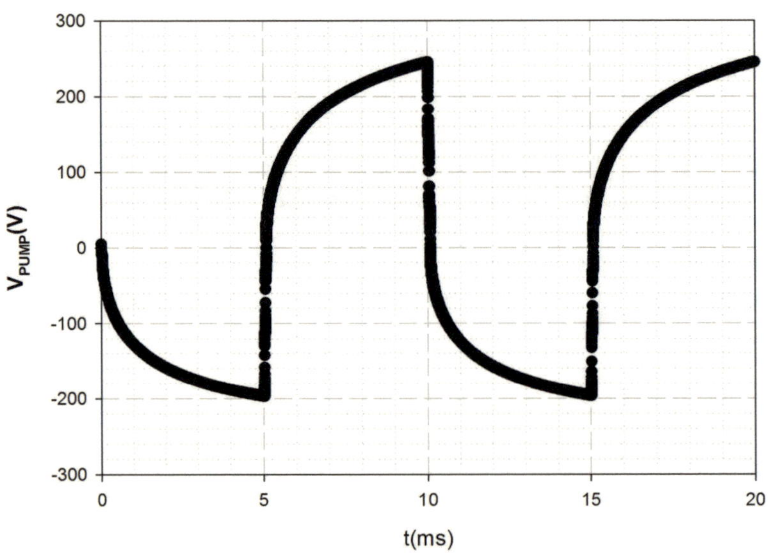

Figure 3.
Micropump excitation voltage simulation.

This discrepancy between positive and negative excitation voltage can be roughly adjusted using different base resistors R_3 and R_4. On the other hand, micropump drivers must be able to independently set the amplitude of micropump excitation signal during positive and negative half-cycle. Analyzed high-voltage circuit, presented in **Figure 1**, does not have such ability. However, due to simple high-voltage circuit topology, the possibility of individual half-cycle amplitude setting was investigated further.

2.2 Large-scale prototype

A large-scale prototype, depicted in **Figure 4**, was designed and implemented. This design was based on through-hole elements for purpose of verification of simulated circuits and selection of appropriate switching transistors.

Switching transistors, which were used during simulations (ZTX657/757), were not available in surface mount housing, therefore other complementary high-voltage bipolar transistors FMMT458/FMMT558 were tested. Better transistor-to-transistor matching and hence better DC offset control was achieved using PBHV9040/PBHV8540 pair of transistors. A test setup, comprised of three Agilent 33120B arbitrary waveform generators and a Siglent SDS1102X oscilloscope was connected to initial prototype: First Agilent 33120B generated micropump operation frequency by supplying square-wave signal to the 40,257 multiplexer, while the other two generators provided square-wave PWM modulated frequencies for high switching frequency excitation of corresponding transistor Q_3/Q_4 (refer to **Figure 1**). Synthesized driving signal frequency was set from 100 Hz to 400 Hz in 100 Hz steps. At each micropump frequency setting, high-switching frequency and duty-cycle were altered with aim to determine output voltage peak-to-peak maximum, whilst maintaining the minimum of DC offset. After each completed micropump frequency sweep, power supply voltage was altered (5 V, 10 V and 12 V). Resulting excitation signal frequency scan vs. micropump amplitude is depicted in **Figure 5**. Amplitude values were very promising, especially in the upper-frequency range (i.e. 400 Hz), where amplitudes up to 240

Figure 4.
Large-scale driver prototype (mutually exclusive SMPS).

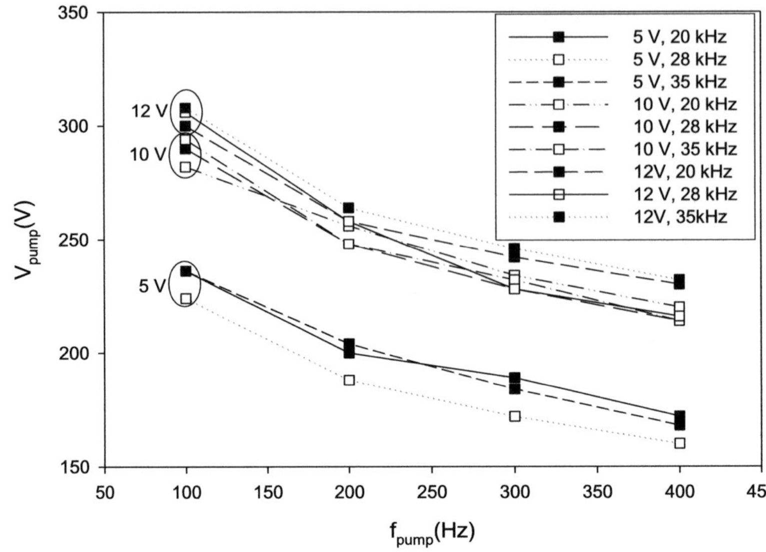

Figure 5.
Frequency sweep of the initial prototype.

V_{pp} were achieved with 12 V power supply voltage. In addition, **Figure 5** shows the boundaries of application range in terms of power supply voltage and high-switching frequency for micropumps, developed in our laboratory [7].

PWM duty cycles of Q_3 and Q_4 transistors, referred to as D_{Q4} and D_{Q3}, which provide maximal micropump amplitude values in **Figure 5**, are summarized in **Tables 1** and **2**. The values are referring to an active part of duty cycle, where a certain transistor is open (i.e. NPN_DRIVE set to V_{CC}, PNP_DRIVE set to GND).

Values, listed in **Tables 1** and **2**, were used in software as pre-programmed duty-cycles, which yield symmetrical positive and negative amplitude of micropump driving signal.

2.3 Module optimization

After assessment of large-scale prototype measurements, presented in **Figure 5**, the true complexity of Q_3 and Q_4 transistors driving signals were revealed–although the circuit operation in **Figure 1** may appear simple, it needs a fairly complex excitation. In order to balance out positive and negative half-period amplitudes in terms of

f_{PWM}(Hz)	20 kHz		28 kHz		35 kHz	
f(Hz)	D_{Q4}	D_{Q3}	D_{Q4}	D_{Q3}	D_{Q4}	D_{Q3}
100	41%	58%	23%	73%	34%	51%
200	40%	62%	26%	73%	41%	56%
300	39%	62%	29%	73%	37%	59%
400	42%	69%	32%	73%	33%	63%

Table 1.
PWM duty cycles (%) for Q_3/Q_4 transistors @ V_{CC} = 5 V.

f_{PWM}(Hz)	20 kHz		28 kHz		35 kHz	
f(Hz)	D_{Q4}	D_{Q3}	D_{Q4}	D_{Q3}	D_{Q4}	D_{Q3}
100	9%	93%	8%	93%	9%	91%
200	9%	94%	10%	94%	9%	93%
300	9%	94%	11%	95%	7%	93%
400	6%	94%	10%	95%	7%	93%

Table 2.
PWM duty cycles (%) for Q3/Q4 transistors @ VCC = 12 V.

DC offset, a single, switched PWM generator, with a switching frequency in order of 30 kHz with at least 7-bit PWM resolution had to be implemented as a cost-effective solution. Micropump output voltage should be monitored and included in closed-loop regulation with adjustment of duty cycle to minimize the driving signal DC offset.

The above-listed requirements would preferably have to be implemented in software, using an 8-bit microcontroller. In our cost-effective implementation, depicted in **Figure 6**, a Microchip ATtiny10 [8], was selected for its price and availability in a 6-pin SOT-23 package. In order to extend the module power supply operating range to 18 V, the microcontroller is connected to a non-inverting, 2-channel MOSFET driver TC4427, which also acts as a voltage level translator for driving signals NPN_DRIVE and PNP_DRIVE (refer to **Figure 1**). Driver inputs are 5 V compatible, therefore the microcontroller can operate at a 3.3 V power supply, thus keeping total current consumption at a minimum, which is a prerequisite for

Figure 6.
Miniaturized driver based on microcontroller ATtiny10.

autonomous (e.g. battery-powered) application. Low voltage 3.3 V microcontroller power supply was generated using LM1117–3.3 circuit.

In order to provide driving signals for Q_3/Q_4 transistors, an 8-bit PWM unit zero in ATtiny10 with two output compare channels (A and B) was used in phase non-aligned mode. Compare value of PWM channel A and B was set using dedicated output compare registers OC0A/OC0B, respectively. Interrupt, which can be triggered upon output compare match with corresponding compare register (OC0A/OC0B), was not used. Instead, each PWM channel toggles between a switching (PWM) state and an inactive (fully open) state. Switching between both states, previously achieved with a '257 multiplexer, can be implemented without additional components, by switching the corresponding pin mode from PWM state (i.e. OC0x mode) to normal I/O state (i.e. PORTB I/O mode).

Achieving switching frequencies in the range of 30 kHz with 8-bit resolution requires the internal oscillator to be configured at 8 MHz. Using such an internal clock signal with no prescaling on timer TMR0 synthesizes a 31.25 kHz PWM clock. Resulting PWM unit features an 8-bit resolution.

Micropump excitation frequency was determined by counting TMR0 timer overflows. As the number of overflows reaches a predefined value, the Q_3/Q_4 transistors excitation roles are reversed by toggling the PWM pin mode and resetting the overflow counter. All this was accommodated inside interrupt routines, without the need for a complex program. In order to compensate for supply voltage variations, the microcontroller also features an analogue-to-digital (A/D) converter, which is used for module power supply measurement. Using A/D readout, PWM driving frequency and both duty cycles were adjusted according to the values, predefined in a software table, derived from **Tables 1** and **2**. This table also encompasses over- and under-power supply voltage detection, which causes the module to turn both Q_3 and Q_4 outputs off. Presented approach implements an open-loop control, which adjusts the duty cycle value according to the power supply voltage.

2.4 Electrical and fluidic characterization

Previously presented optimized module was tested for current consumption and output signal accuracy with different capacitive loads connected to the output. A multimeter on the power supply line was used to measure current consumption and an oscilloscope was used to measure output signal parameters (waveform shape, amplitude symmetry, frequency). High voltage variable capacitor with values from 10 nF to 100 nF and several piezoelectric micropumps in parallel were used to vary the capacitance of output load. A constant micropump operation frequency was set at 190 Hz, while the power supply voltage was set to 9 V.

Though the resulting capacitance driving capability in **Figure 7** exhibits practically linear dependency, it must be emphasized, that the signal shape deteriorates severely and the positive signal slew rate drops from 140 V/ms range to 80 V/ms.

Measured limiting value of load capacitance, which maintains 140 V/ms slew rate was found at 57 nF with a driving voltage of 172 V_{pp}. Larger capacitive loads (> 57 nF) cause the micropump driver current to saturate, which deteriorates the output signal rise- and fall-time and correspondingly the slew-rate. This reduction consequentially results in a drastic deterioration of micropump flow and backpressure characteristics. Such output signal deterioration was attributed primarily to limited current passing through the inductor L_1.

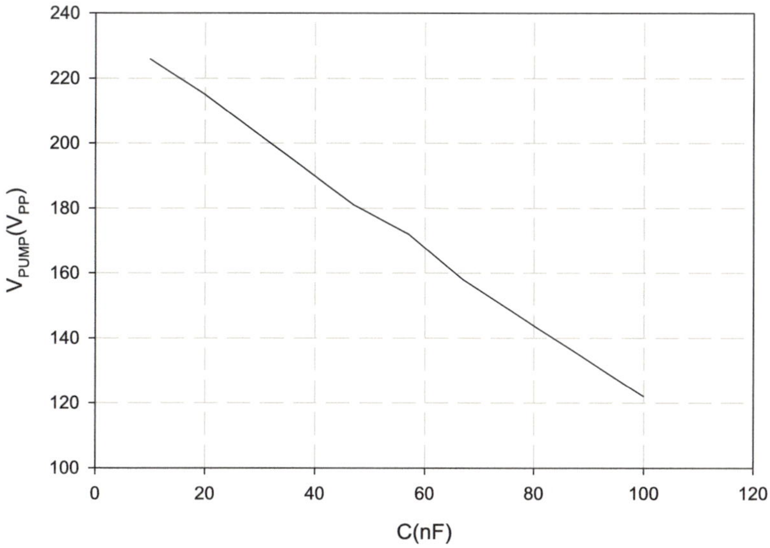

Figure 7.
Micropump driving voltage vs. variable load capacitance.

In order to achieve better signal integrity, peak L_1 inductor current should be increased–which is in contradiction with low-power design aspect. Due to unavailability of higher-quality inductors, all our designs used a Bourns SRR0603-102KL inductor [].

Figure 8 represents measured results of output signal with 12 nF capacitive loading, achieving 236 V_{PP} drive amplitude, with positive and negative slew rates 142 V/ms and 163 V/ms, respectively. Positive cycle amplitude was 122 V, while the negative was −114 V.

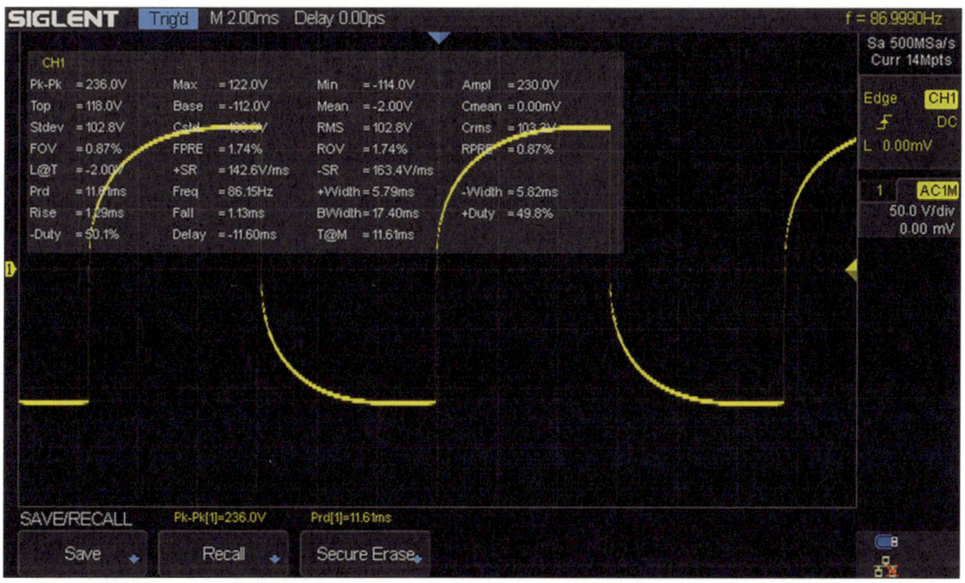

Figure 8.
Micropump driving voltage at 87 Hz.

Afterwards, the module current consumption was measured after performing a sweep of power supply voltages.

Driver current consumption vs. power supply voltage is depicted in **Figure 9**. Module current consumption increases linearly with respect to increasing load and signal frequency up to 55 mA at 12 V power supply, which was the upper design interval

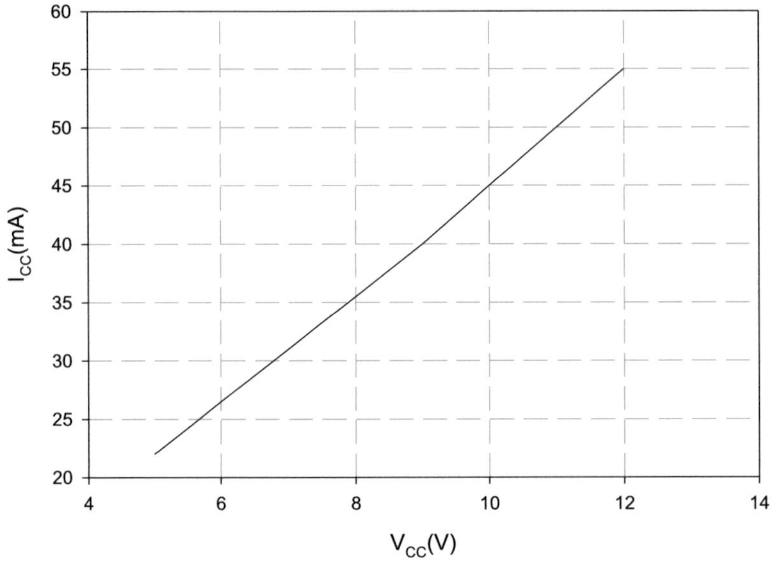

Figure 9.
Driver current consumption.

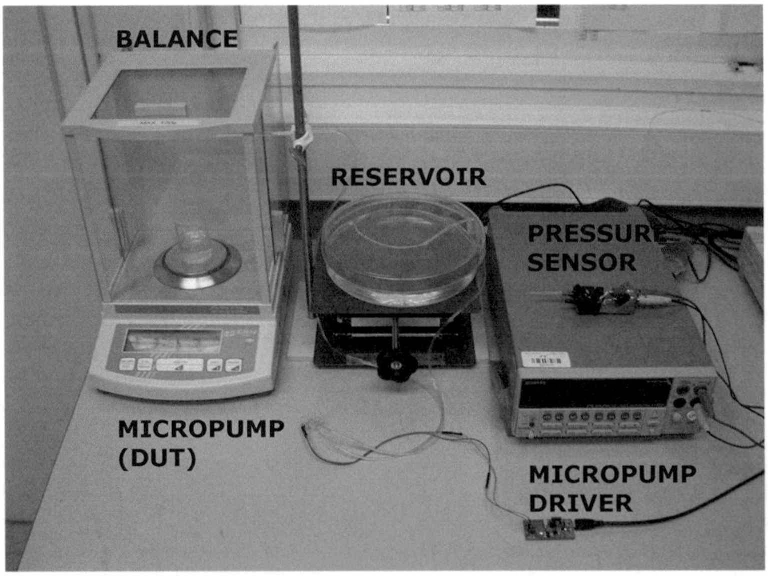

Figure 10.
Measurement setup for micropump fluidic characterization.

limit and also the power dissipation limit of the driver TC4427. Microcontroller current consumption remains negligible throughout entire measurement range. Measured driver current consumption was found between 22 mA (5 V_{CC}) and 55 mA (12 V_{CC}).

A dedicated computer-controlled system for characterization of piezoelectric micropumps was set up after electrical evaluation of the optimized micropump driver. The measurement system, presented in **Figure 10**, enables frequency scans and allows simultaneous pressure and flow measurements.

Analyzed micropump driver was connected to a micropump, developed in our laboratory [7]. A reservoir, filled with DI water, was connected to the micropump input and pressure/flow evaluation equipment on its output.

Micropump flow was measured indirectly by weighing the mass of pumped media over a known period of time using a Kern ABJ 120–4M precision scale. Obtained results were corrected to account for the evaporation of medium during the measurement, which was determined in separate experiments.

Flow measurements are presented in **Figure 11**. Results show frequency scan with comparison to Bartels mp-x Controller [1], which was preset with the same amplitude and frequency conditions. The only difference in driving parameters was the signal shape, where the Bartels mp-x Controller was set to square-wave signal, and the tested driver provided RC like shape, presented in **Figure 8**. The same micropump was driven on a frequency interval from 50 Hz to 400 Hz and both duty cycles were set according to values in **Tables 1** and **2** to achieve minimal DC offset of drive voltage.

Figure 12 summarizes obtained both positive (SR^+) and negative (SR^-) slew-rates as well as micropump driving voltage amplitude (not peak-to-peak) with respect to frequency.

For lower frequency range (i.e. under 80 Hz), designed driver actually supersedes the flow performance of Bartels mp-x controller, achieving 1.2 ml/min in comparison to 0.8 ml/min obtained with mp-x module. However, as the driving frequency increases, the slew rate and drive amplitude decrease, resulting in a steady exponential decay of flow performance. As can be observed from **Figures 11** and **12**, best flow

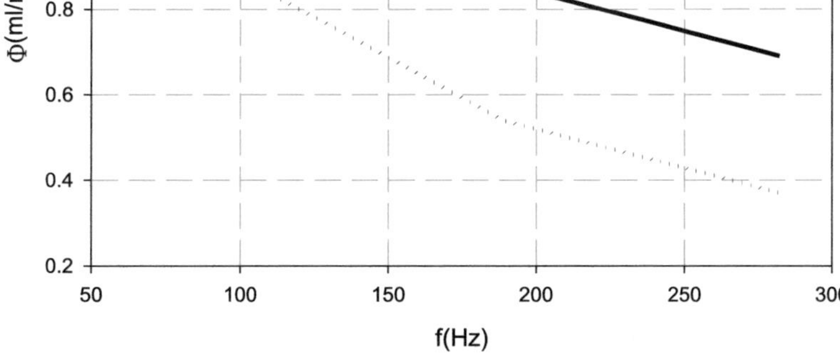

Figure 11.
Micropump DI water flow rate vs. frequency.

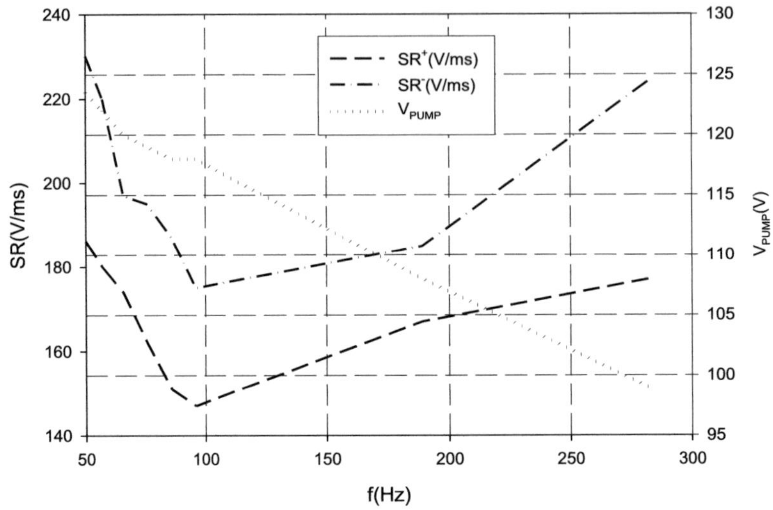

Figure 12.
Slew rate/drive amplitude vs. frequency.

performance of the designed driver is achieved at low frequencies (< 80 Hz), where it is primarily limited by driving voltage amplitude V_{PUMP} and both slew-rates SR^+ and SR^-, depicted in **Figure 12.**

2.5 Influence of excitation signal asymmetry on micropump performance

It is generally accepted, that the shape of the driving signal affects the operation of micropumps. Typically, sine- or square-wave micropump driving signals with symmetric positive and negative amplitudes are most often applied [9–12].

In order to investigate the effect of amplitude asymmetry on micropump performance, driving signals of various amplitude asymmetries had to be synthesized. Since the synthesized signal amplitude depends not only on the output load impedance but also on the duration of each half-cycle by extending the time of first and shortening the time of the second half-cycle, the synthesized signal amplitudes increase and decrease, respectively. In other words, if one boost converter has more time available than the other, it will build up comparatively higher voltage on piezoelectric actuator. The time of both half-cycles is defined by a duty cycle setting. This parameter can be arbitrarily set due to the flexibility of the built-in microcontroller.

We introduced a unique approach, which employs an adjustment of the duty cycle ranging from 30–70%, enabled synthetization of excitation signals with various degrees of amplitude asymmetry, needed for elastomeric micropumps flowrate and backpressure performance characterization. The degree of signal asymmetry is defined as $S_{ASIM} = V^+ - V^-$. Synthesized signal for four representative duty cycle settings (DC^+ = 30%, 40%, 50% and 60%) and for the duration of 22 ms when driving piezoelectric actuator P-5H (Sunnytec Suzhou Electronics Co., Ltd. [13]) is shown in **Figure 13**. When the duty cycle was set to DC^+ = 30%, the positive amplitude V^+ was lower than the negative amplitude in spite of higher signal slew rate in the first half-cycle. At duty cycle, DC^+ = 40% amplitude symmetry of driving signal is reached. The maximal value of the applied electric field is 600 V/mm.

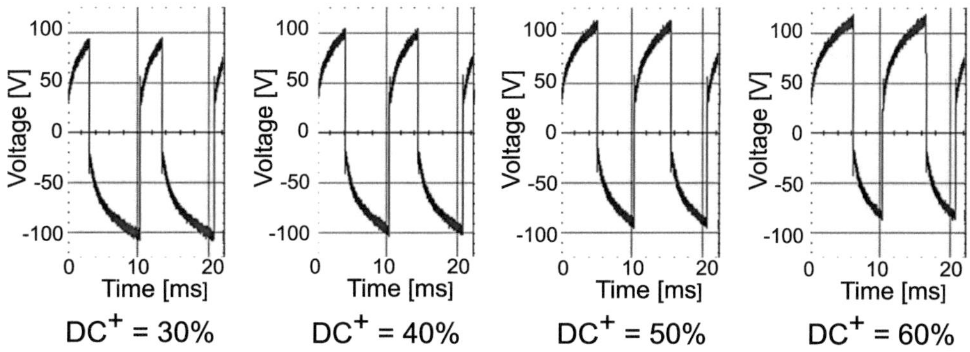

Figure 13.
Synthesized excitation signal for four representative values of duty cycle for the duration of 22 ms.

Figure 14 shows micropump [14] flowrate and backpressure performance characteristics vs. excitation signal duty cycle ranging from 30 to 70% at a constant excitation frequency of 100 Hz. Module power supply voltage was set to 9 V.

Excitation signal amplitudes V^+ and V^-, flowrate and backpressure performance for DI water medium were measured for each of the duty cycle setting using the measurement setup, described in **Figure 10**. By extending the DC^+ duty cycle, the amplitude of the positive half-cycle increases while the negative one decreases. The characteristics are similar to linear functions with opposite slopes $+0.43\ V\%^{-1}$ and $-\ 0.43\ V\%^{-1}$. Measured characteristics show that flowrate and backpressure performance values are the highest at the duty cycle of equalization DC_{eq}^+, which is 40%. We assume that in this setting, the efficiency of the suction stroke most probably equals the efficiency of the compression stroke.

By extending the DC^+ duty cycle beyond 40%, the excitation signal amplitude V^+ during micropump compression stroke increases and the amplitude V^- during the suction stroke decreases.

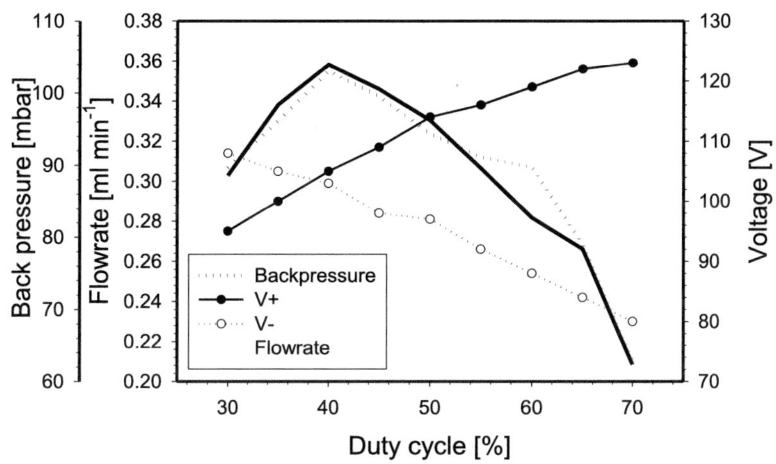

Figure 14.
Micropump flowrate and backpressure performance characteristics vs. excitation signal duty cycle ranging from 30 to 70% at a constant excitation frequency of 100 Hz.

Remarkably, the comparatively higher excitation signal amplitude V^+ in the compression stroke fails to compensate for the decrease in the amplitude V^- in the suction stroke and the pumping performance declines. The same trend of performance decline is observed when duty cycle DC^+ is set below 40%. The higher excitation signal amplitude V^- in the micropump suction stroke cannot compensate for the decrease in the amplitude V^+ in the compression stroke and the pumping performance declines again.

Flowrate and backpressure performance of micropump for DI water decreases rather exponentially with increasing degree of signal asymmetry. Based on these characteristics results, it is presumed that the suction and compression stroke performance must be balanced for the highest micropump flowrate and backpressure performance. Hence, the total performance of the pump might only be as high as is dictated by the performance of less efficient stroke. This exponential decrease of pumping performance vs. increasing degree of signal asymmetry coincides with the exponential increase of the pumping performance vs. increasing excitation signal amplitude [15]. Above mentioned exponential trend results from employed active rectifying elements (sequential expansion and throttling of rectifying elements are performed by actuated membrane deformation). If the excitation signal amplitude is reduced, micropump membrane deformation decreases. On the other hand, micropump displacement volume and active rectifying elements efficiency are reduced also.

For micropumps employing passive check valves (sequential opening and closing of check valves are performed by fluidic flow), flowrate and backpressure performance vs. excitation signal amplitude are linear [15]. As the excitation signal amplitude decreases, the displacement volume of micropump decreases, while the efficiency of the passive check valves remains constant. It is speculated that the flowrate and backpressure performance of micropumps employing passive check valves would decrease proportionally with the degree of excitation signal asymmetry S_{ASIM}. It was shown, that micropump flowrate and backpressure performance is limited by a lower amplitude when driven by amplitude-asymmetric signal and that for maximum micropump performance the amplitude symmetric signal is required. Indeed, the performance degradation due to excitation signal asymmetry might be compensated by increasing both the negative and positive half-cycle voltage V^- and V^+ e.g. by increasing the controller's power supply voltage. However, with such an approach, the maximum permissible electric field in the piezoelectric actuator would be exceeded, which equals only 490 V/mm for the PZT-5H standard composition [13]. This would result in permanent actuator performance degradation due to depolarization effect. From measured results (**Figure 14**), it can be concluded that micropumps should be driven by an amplitude-symmetric excitation signal. Therefore, it is mandatory to introduce a solution for providing amplitude symmetry to ensure stable long-term micropump operation and high backpressure and flowrate performance characteristics.

2.6 Impact of excitation signal amplitude symmetry on micropump performance

To compensate for the influence of piezoelectric load and to balance the signal, we introduced a solution based on fine-tuning of duty cycle setting. By prolonging the time of first and shortening the time of the second half-cycle, an amplitude-symmetric signal can be synthesized despite piezoelectric load. The effectiveness of the solution was demonstrated on PZT sample P-5H. The microcontroller code for module was modified to enable adjustment of 16 different signal/pause signal ratios from 20 to 80% in the frequency range from 90 to 150 Hz by 10 Hz increments. The power supply voltage for control module was set to 9 and 10 V, respectively. The module was programmed to a chosen excitation frequency and for each duty cycle setting, amplitude of positive half-cycle V^+, of negative half-cycle V^- and power supply current I_{CC} were measured. **Figure 15** shows the degree of signal asymmetry S_{ASIM} as a function of the power supply voltage and duty cycle at excitation frequencies from 70 to 150 Hz. We observed that duty cycle of equalization DC_{eq}^+ at 9 V and 10 V of power supply voltage ranged between 30 and 38% and between 40 and 43%, respectively.

Figure 16 shows the driving signal amplitude as a function of duty cycle DC^+, frequency *f* ranging from 70 to 150 Hz and power supply voltages of 9 and 10 V. From **Figure 16**, it follows that the maximum amplitude (120 to 130 V) is achieved only in the interval up to 100 Hz (curves 125 and 120 in **Figure 16**, left), while this limit moves up to 110 Hz at higher power supply voltage (10 V, curves 130, 125 and 120 in

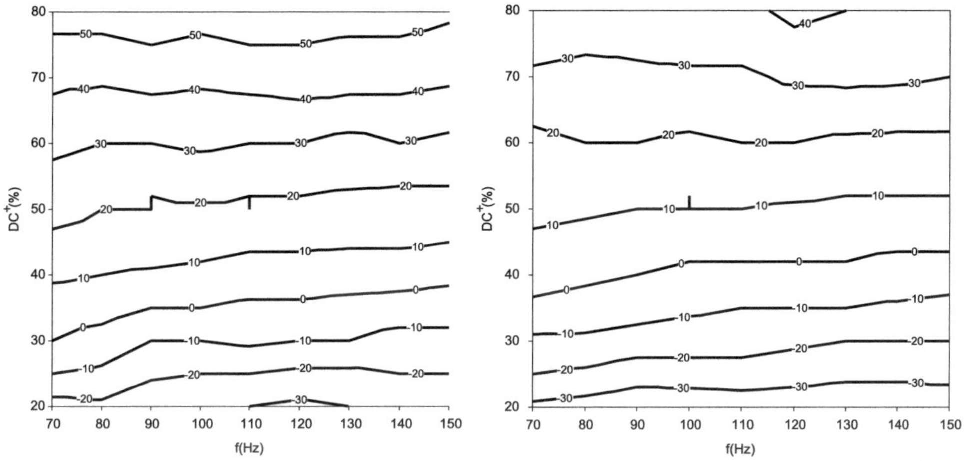

Figure 15.
The degree of signal asymmetry SASIM as a function of duty cycle DC⁺ at excitation frequencies f ranging from 70 to 150 Hz at 9 V (left) and 10 V (right) power supply voltage.

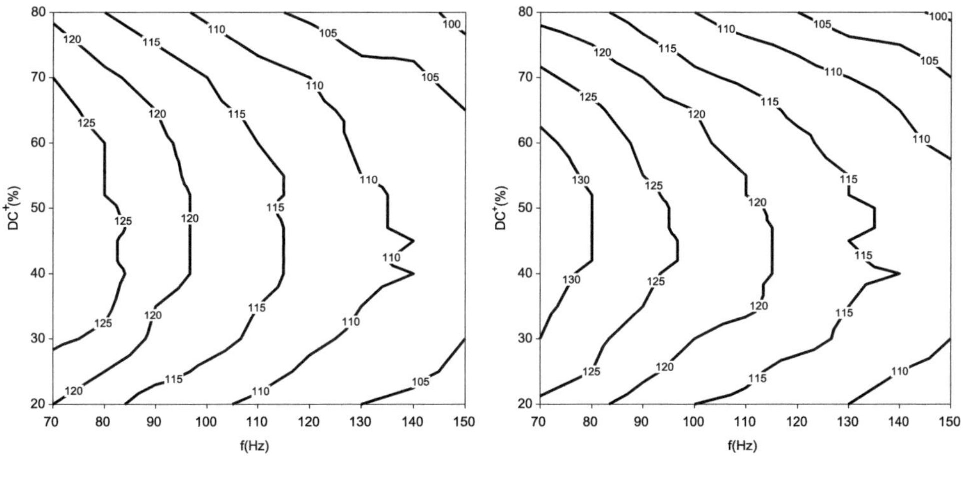

Figure 16.
Micropump driving signal amplitude in V as a function of operating frequency and duty cycle DC⁺ for power supply voltage of 9 V (left) and 10 V (right).

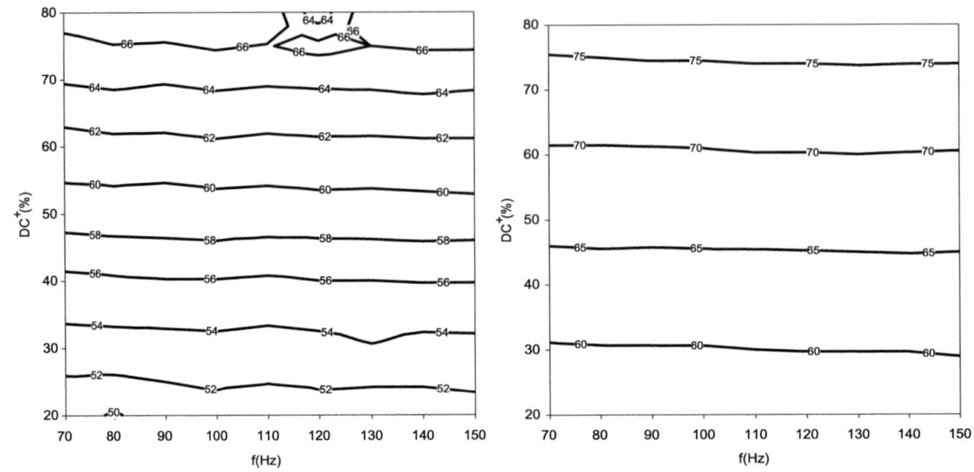

Figure 17.
Power supply current in mA as a function of operating frequency and duty cycle DC⁺ for power supply voltage of 9 V (left) and 10 V (right).

Figure 16, right). A higher excitation frequency (Δf = +10%) affects the flow rate and backpressure performance of piezoelectric micropumps.

Figure 17 shows the power supply current as a function of the power supply voltage and the duty cycle DC^+ at various frequencies ranging from 70 to 150 Hz. From **Figure 17**, it follows that the power supply current is not affected by module operating frequency, which is also supported by the theory of switching power supplies but increases almost linearly with duty cycle DC^+ setting.

The upper limit of duty cycle setting, which keeps the power supply current below 70 mA is 75% and 62% for 9 V and 10 V of power supply voltage, respectively.

Signals of varying degrees of amplitude asymmetry were synthesized by varying the duty cycle setting. Backpressure and flowrate performance of the micropump was shown to be the highest at the duty cycle of equalization DC_{eq}^+ = 40%. To compensate for piezoelectric load and to provide amplitude symmetric driving signal, a solution by fine-tuning of duty cycle setting was proposed.

Duty cycle of equalization DC_{eq}^+ at 9 V power supply voltage ranged between 30 and 38% while at 10 V power supply voltage shifted toward 43%.

3. Optocoupler-based micropump module design

Design, implementation and characterization of a miniaturized piezoelectric micropump driving module, based on two boost SMPS converters, with a shared boost capacitor (micropump PZT actuator) represented a starting point in design of micropump drivers in our Laboratory. This micropump driver design is suitable for integration inside micropump housing due to its small size and its low current consumption (≈ 55 mA).

Such design synthesizes rectangular shape of driving signal with RC charging and discharging transitions, which are not optimal. Low value of both transition slew rates does result in lower current consumption, but at a cost of reduced pump flow rate compared to a steeper slew rate signal. Still, the performance of such a circuit can be considered suitable for certain cost- and size-sensitive applications.

Fluidic measurements have shown the need for assuring driving signal symmetry (i.e. elimination of difference between positive and negative amplitude), which has to be trimmed individually to a particular type of piezoelectric actuator, which was another drawback of this design.

Module upper-frequency limit was found to be at 400 Hz. A typical operation range was found up to 100 Hz, where peak-to-peak amplitudes up to 250 V could be achieved. For DI water pumping, optimal operating frequency range was found between 50 to 80 Hz. Amplitude symmetrical driving signal with an amplitude of 125 V was achieved using power-supply voltage of 9 V, excitation frequency f of 100 Hz and duty cycle DC^{+} of 40%. Using these optimal driving conditions considering driving module current consumption and micropump performance characteristics, micropump flowrate performance was found at 0.36 ml min^{-1}, backpressure performance of 104 bar and module power consumption of 0.5 W.

Based on presented fluidic and electrical measurement results, obtained rectangular shape of driving signal with RC charging and discharging transitions, had to be improved to allow faster signal edge transitions (i.e. improved slew rate). Future driver designs also had to allow for a more precise (preferably independent) setting of positive and negative amplitude. Lastly, a wider frequency operation interval than 400 Hz would have to be achieved in order to improve air pumping characteristics.

To implement a driving signal shape of a square wave with settable frequency, duty cycle, both amplitudes and both slew rates, a completely different approach to the design of a miniature high voltage piezoelectric micropump driving module was taken: such design had to synthesize a proper rectangular micropump driving signal with independently settable positive and negative amplitudes, rising and falling edge slew-rates, positive and negative dead-times and excitation frequency.

In order to achieve these functionalities, a simplified high-voltage driving stage was designed. Typical approach would be based on a D-class amplifier design using an isolated gate driver (e.g. STGAP series), which would enable driving piezoelectric micropumps with arbitrary waveforms. On the other hand, such an approach requires several additional components, which would compromise our low-cost approach.

In order to reduce overall cost, the micropump high-voltage switching stage was highly simplified. While isolated gate drivers do provide galvanic isolation between the input section (a microcontroller) and output transistors, they also require a bootstrapping capacitor for high-side switch. Optocouplers perform the necessary level-translation for the high-side transistor and eliminate the need for any additional bootstrap capacitors. Optocoupler-based solution also enables fast transitions and allows for a settable slew-rate as will be presented in the following.

3.1 Module operation

Optocoupler-based design comprises a separate negative and positive high-voltage power supply, each implemented as an independent boost SMPS power supplies, which generate positive and negative micropump driving voltage (**Figure 18**, netlabel V^{+} and V^{-}) from a low voltage source (**Figure 18**, VPWR).

Positive micropump power supply voltage generator is comprised of transistor M1, inductance L1, diode D1 and capacitor C1, while the negative micropump power supply voltage comprises transistor M2, inductance L2, diode D3 and capacitor C3. Both power supply voltage levels (**Figure 18**, netlabel V^{+} and V^{-}) are independently monitored using a dedicated resistor divider (**Figure 18**, resistors R4, R5 and R7, R8 for negative and positive voltage, respectively). Each resistor divider forms a

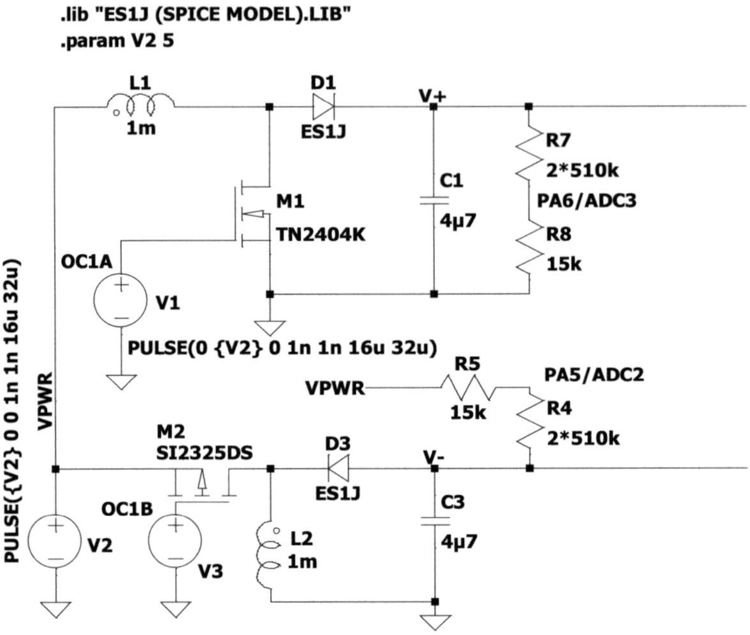

Figure 18.
High-voltage part of the optocoupler-based driver design.

closed-loop voltage regulator with a corresponding microcontroller PWM source (**Figure 18**, labels V3 and V1 for negative and positive voltage, respectively).

Both SMPS converters are driven using PWM signals on sources V1 and V2 (**Figure 18**) with a base frequency of 32 kHz. Achievable duty cycle range is from 10–90%. Depending on duty cycle setting, both SMPS converters can independently deliver output voltage in a range from ±10 V to ±150 V. Independent setting of both positive and negative power supply voltage allows the synthesis of a 50% time-symmetrical signal. This could not be achieved using previously presented design, where amplitude symmetry could only be achieved by adjusting the duty cycle of micropump excitation period.

High-voltage switching stage comprises two Darlington output high-voltage optocouplers Toshiba TLP187 (**Figure 19**, circuit U1 and U2). Positive power supply voltage (**Figure 18**, label V^+) is connected to positive voltage switching optocoupler (**Figure 19**, circuit U1), while the negative power supply voltage (**Figure 18**, label V^-) is connected to negative voltage switching optocoupler (**Figure 19**, circuit U2). Independent setting of both power supply voltages and opening of positive and negative voltage optocoupler enable independent setting of rising and falling edge of driving signal as well as positive/negative duty cycle and consequently excitation frequency. Such independent driving introduces an additional part of micropump operating positive/negative half-cycle, called positive/negative dead time, during which both optocouplers are turned off.

In **Figure 19**, only resistors SR1 and SR2 are shown to limit complexity, but more microcontroller output pins may be connected to optocoupler diode input, each via its different value resistor. Using such a setup, a simple D/A (digital-to-analog) converter is formed, which can be used for digitally setting the value of rising/falling edge slew-rate. To enable more fine setting of both slew rates, a microcontroller with an onboard D/A could be used.

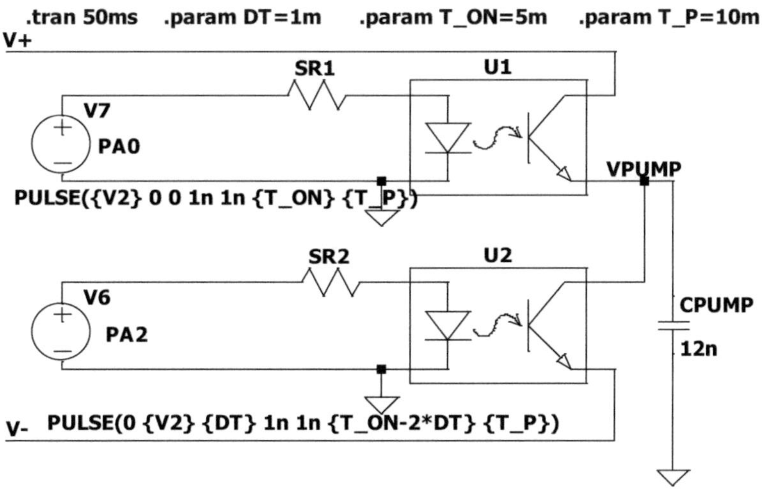

Figure 19.
Excitation part of the optocoupler-based driver design.

Rectangular micropump excitation signal is synthesized by providing phase-synchronized optocoupler driving signals V6 and V7 (see **Figure 19**, **Figure 20**). Both driving signals must include a programmable dead-time gap, which is downwards limited by optocoupler turn-off-time (80 μs for TLP187). Any further reduction of this minimal dead-time gap value would shorten out both power supplies.

The micropump switching period is divided into four stages as follows: Positive driving micropump voltage stage, noted by a solid line in **Figure 20**, where positive power supply (**Figure 18**) is connected to the micropump via optocoupler U1 (**Figure 19**). Subsequently, a positive dead time state is next, where both optocouplers are turned off. Following positive dead-time stage, there is a negative driving micropump voltage stage, noted by a dashed line in **Figure 20**. During this stage, negative SMPS power supply (**Figure 18**) is connected to the micropump using

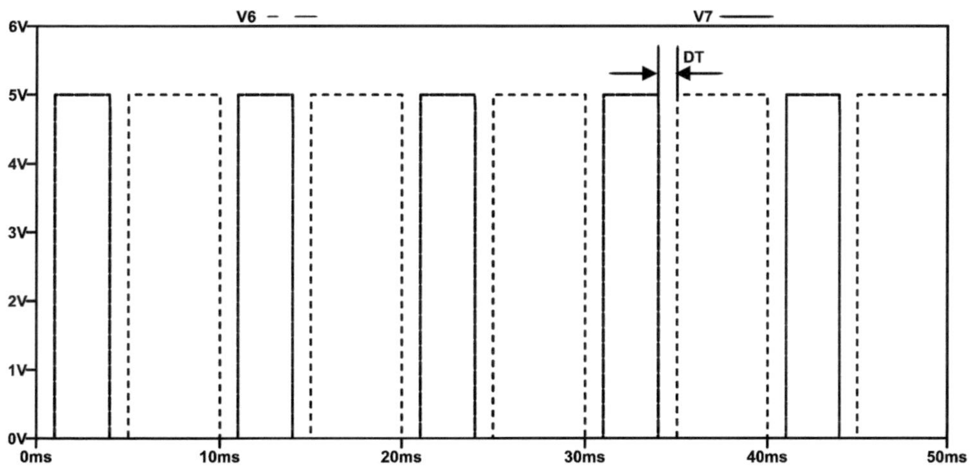

Figure 20.
Optocoupler diode driving voltage.

optocoupler U2 (**Figure 19**). Micropump excitation period finishes with a negative dead-time stage, where both optocouplers (U1, U2) are again switched off.

Such division of micropump excitation period into four independent stages yields maximum control over micropump driving signal parameters and overall current consumption. Decreasing the value of optocoupler based resistors SR1, SR2 (**Figure 19**) results in higher micropump driving signal slew rates. Selection of resistors represents a compromise between high-flow performance, which is achieved by fast slew-rate and consequentially higher current consumption, and low-flow performance, which is achieved by lower slew-rate and lower current consumption.

Adjusting optocoupler base current using a simplified D/A converter by connecting different resistors enables seamless interchanging between performance modes and different driving signal shapes: Sinus-flank signal (sinusoidal waveform with prolonged flat maximum/minimum) can be obtained by setting both rising and falling slew-rate low using a single, high-value resistor is fed to optocoupler diode. Rectangular signal can be synthesized by setting both slew rates high by turning on multiple resistors. Sinewave-rectangular signal (SRS signal) can be obtained by setting rising edge slew-rate high and falling edge slew-rate low.

Setting of both power supply voltages, slew-rates and dead time preferably have to be implemented in software, using an 8-bit microcontroller. In our cost-effective implementation, depicted in **Figure 6**, a Microchip ATtiny104 [16], was selected for its price and availability in a 14-pin SO package. An 8-bit timer zero is used both for counting through four distinctive signal stages and two-channel PWM synthesis for high-voltage generation. Timer zero is counting using an internal 8 MHz oscillator with no Prescaler, which produces two independent (non-aligned) channel (A and B) 8-bit PWM outputs running at 31.25 kHz. Each PWM channel features a corresponding output compare register (OCR0A/OCR0B).

Driving waveform state machine is stepping through active and dead-time stages for positive/negative amplitudes. Transition to the next state is achieved by presetting timer zero expiration period. After this period, timer zero overflow interrupt causes transition to next state and renewed calculation of next-stage period. Waveform transition state machine can be omitted entirely – in this case, constant positive or negative power supply voltage is connected to the micropump, effectively turning the micropump into a valve.

PWM output (OC0A/OC0B) value results from comparison between OCR0x and timer zero value. These PWM outputs are connected to the corresponding switching transistor (M1, M2 in **Figure 18**). The output (V^+, V^- in **Figure 18**) of positive/negative power supply SMPS is being monitored using a resistor divider, which is connected to corresponding microcontroller analogue-to-digital converter (ADC) input (ADC2/ADC3, **Figure 18**). An ADC conversion complete interrupt is enabled by timer zero interrupt routine only during positive and negative dead-time switching phase. During positive dead-time state, ADC2 is monitored, and the value of OCR0A is changed accordingly. During negative dead-time state, ADC3 is sampled, and the value of OCR0B is changed. Both positive/negative monitoring algorithms are based on a fast proportional regulator. Microcontroller UART receiver was used to configure all micropump driving signal parameters: frequency, positive and negative amplitudes, dead times, slew-rates and operation type (pumping/valve). In order to reduce the burden of calculations on a microcontroller side, an Excel VBA based script was developed. This script translates human-readable parameters such as frequency and dead time to timer zero state-machine expiration periods, determines slew-rate resistor multiplexing state and configures the mode of module operation.

Switching optocouplers PC817 (U1, U2 in **Figure 19**), which were used during simulations, were replaced with high-voltage, Darlington output type (Toshiba TLP187), due to better switching characteristics and high-current transfer ratio, which achieves higher slew rates. Large (10 μF/100 V) rectifying capacitors (C1, C3 in **Figure 18**) were initially used to minimize the power supply ripple. Their size was reduced during the following measurements.

A digital storage oscilloscope was connected to the driving module prototype. Tested module was driving the micropump with PZT capacitance 12 nF, developed in our laboratory [7]. Developed module can set output signal dead time, between two extrema:

1. Micropump driving voltage is applied almost for the entire corresponding half-cycle, except for dead time, which is set to DT = 200 μs. This is called full drive operating mode. **Figure 21** depicts the front-edge transition of micropump output signal during full drive operating mode at 100 Hz and 100 V. Achieved micropump power supply amplitudes were ± 98 V.

2. Micropump driving voltage is applied *only* 200 μs of driving time for entire corresponding half-cycle, otherwise signal remains inactive. This is called minimum drive operating mode. Note, that the dead-time (DT) during minimum drive operating mode is depending on the driving signal period (DT = T/2–200 μs). Typical dead time clamping value of 200 μs was selected to allow correct optocoupler switching and microcontroller ADC conversions.

Micropump output signal of initial prototype during minimum drive of 200 μs with driving frequency 100 Hz and amplitude ±100 V is shown in **Figure 22**. Piezoelectric

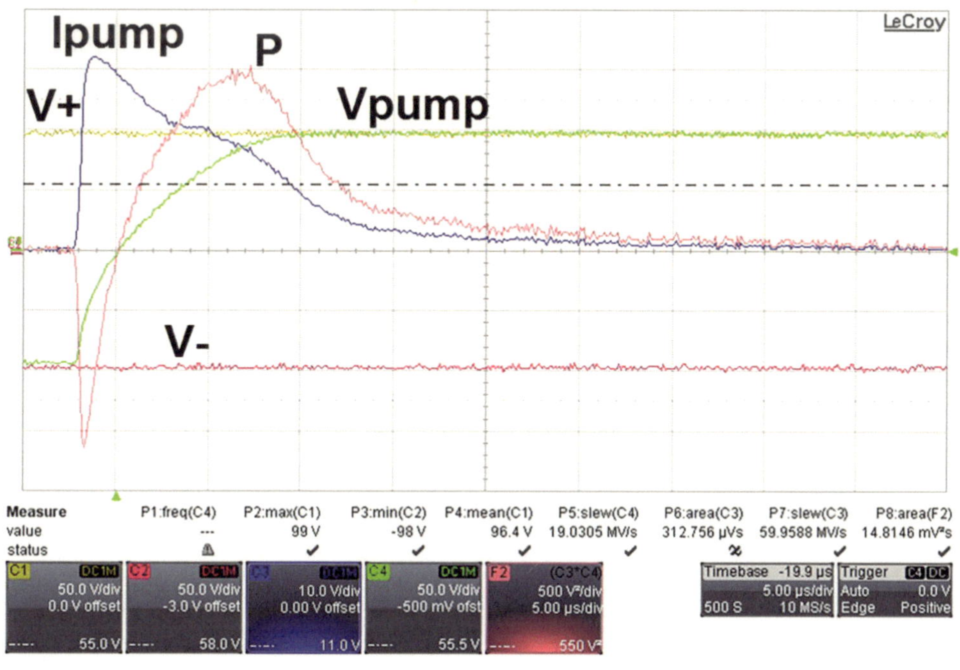

Figure 21.
Initial prototype full drive operation mode front edge detail.

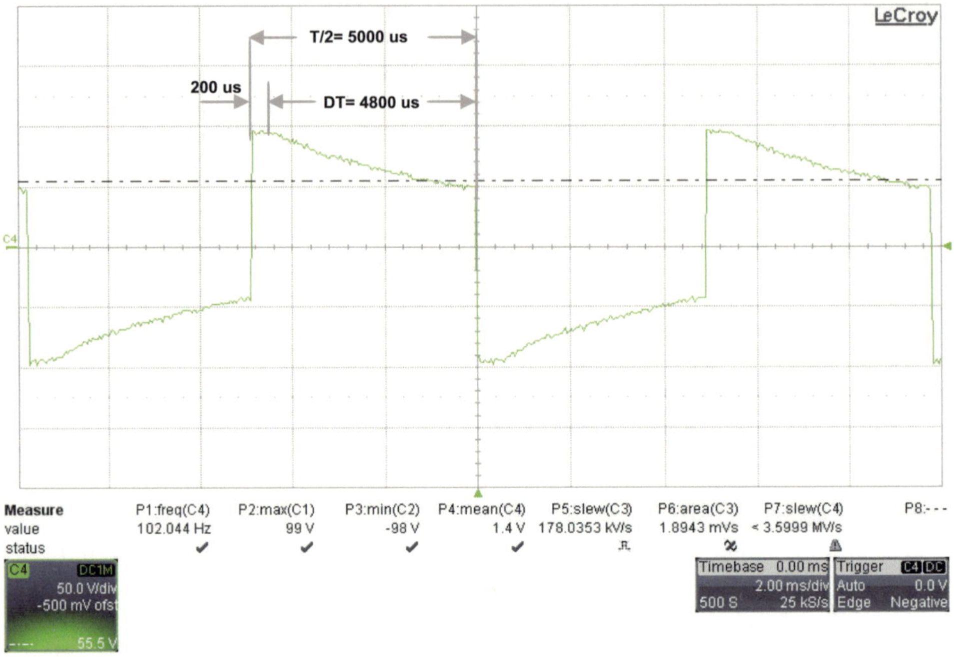

Figure 22.
Initial prototype minimum drive operation mode micropump driving signal.

actuator voltage falls off according to capacitive discharge down to 50 V after initial driving with amplitude ±98 V for 200 μs (max/min section, **Figure 22**). The time constant of capacitive discharge decay is practically independent of voltage polarity, on the other hand, its value changes depending on pumped media viscosity and micropump design.

3.2 Electrical characterization

Driving signal frequency was investigated in range of 50 Hz to 1 kHz for full and minimum drive operating modes. Micropump driver was configured with maximum piezoelectric actuator voltage of 125 V using Excel control software. Maximum actuator voltage is limited by a maximum field of 600 V/mm of piezoelectric actuator P-5H (Sunnytec Suzhou Electronics Co., Ltd. [13]). Current consumption, positive and negative power supply amplitudes, slew rates and were measured at each micropump frequency value. **Figure 23** is presenting obtained results for micropump amplitude at each excitation frequency setting.

Micropump driver was able to reach admissible voltage limit of ±125 V in the frequency range up to 150 Hz, which extended initial clamping interval of 70 Hz, achieved with voltage scan of the initial prototype. This extension consequentially increases module current consumption I_{CC}, which is presented in **Figure 24**.

At the target operating frequency of 100 Hz, in full drive operating mode with $SR^+ = SR^- = 16$ V/μs (**Figure 25**), module current consumption was clamped to 118 mA, at 150 Hz the current consumption increases to 180 mA. In minimum drive operating mode, current consumption reaches 170 mA. Difference in current consumption between 'full' and 'minimum' driving mode is minimal (10 mA), therefore, the majority of current consumption is attributed to decrease in efficiency of SMPS

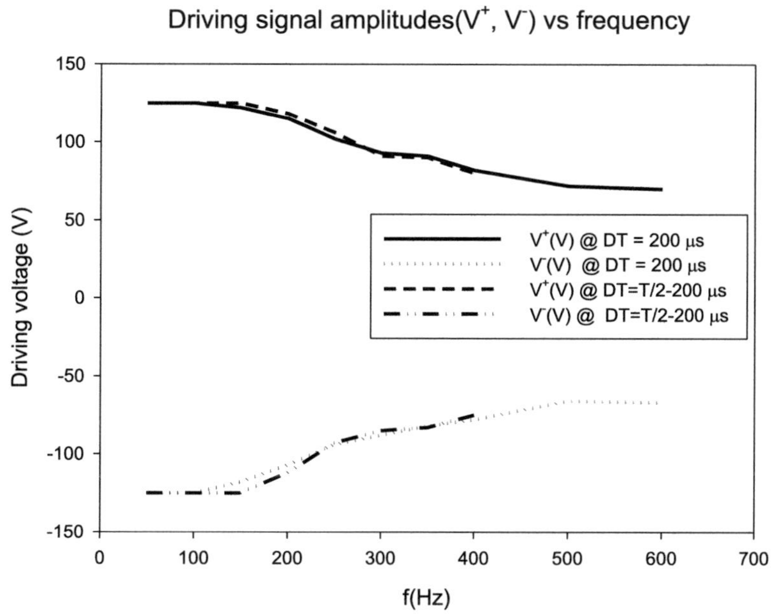

Figure 23.
Voltage-frequency sweep.

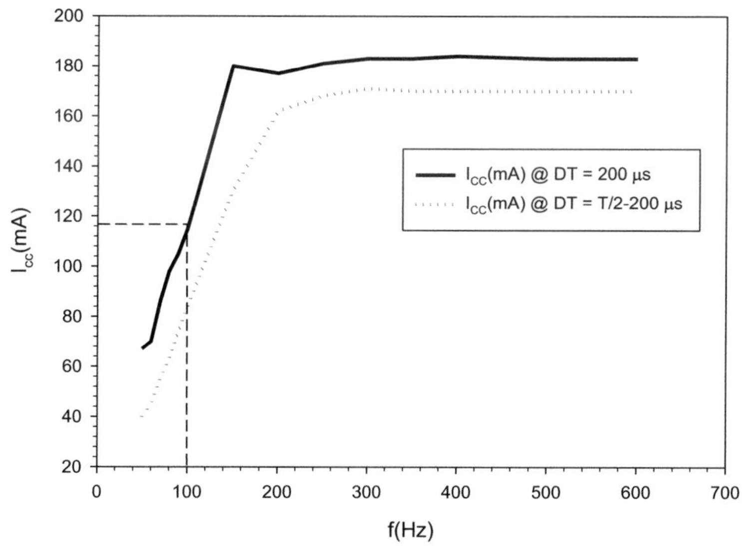

Figure 24.
Current consumption vs. frequency.

boost converters, when trying to supply actual current to the micropump actuator at such maximum admissible voltage setting. SMPS boost converter current is primarily limited by current capabilities of the used SRR0603 inductor. Current limit value of 180 mA could only be extended by raising the power supply value from 9 V to 10.5 V further to 200 mA for a short period of operation.

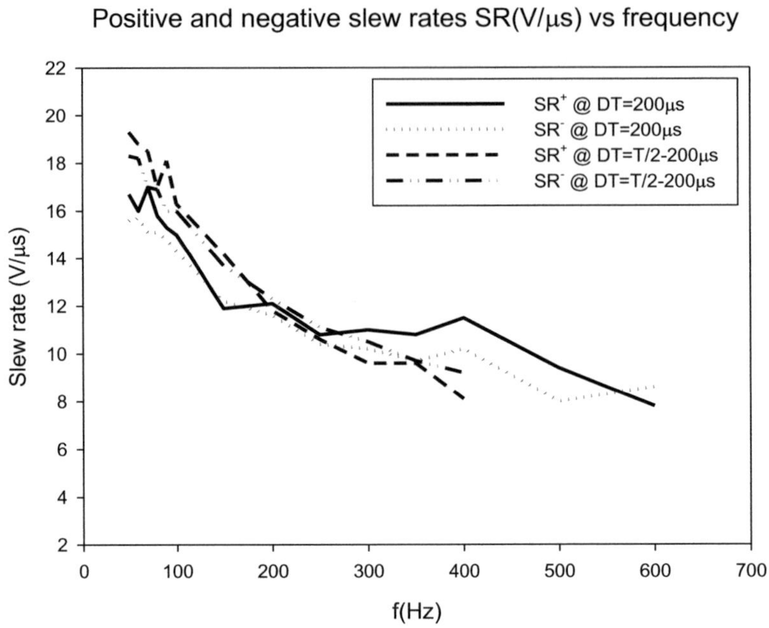

Figure 25.
Positive and negative slew rate vs. frequency.

Extension of frequency interval up to 150 Hz (**Figure 23**) with operation at maximum driving amplitude of ±125 V, resulted in elevation of both slew rates from initial 11 V/μs to 16 V/μs, when configured to full drive operating mode at 100 Hz, as depicted in **Figure 21**. In minimum drive operating mode, this value is even higher (18 V/μs). Slew rate remains well over 10 V/μs in the frequency range to 400 Hz, which enables evaluation of our micropumps, with a smaller (Φ = 6 mm) piezoelectric actuator disc with a capacitance of 4 nF [15].

Next, module current consumption was evaluated against slew rate in the full drive operating mode, with micropump excitation frequency of 100 Hz and amplitudes ±125 V. Measured module current consumption was 118 mA, with both slew-rates set to 16 V/μs, shown as a dashed line **Figure 21**.

Waveform measurements, such as frequency, amplitude, both slew rates and waveform averaging were evaluated instantaneously by the oscilloscope. Microcontroller I/O port output driving capability limited any further lowering of optocoupler base resistance to 220 Ω. On the other hand, high limit of optocoupler base resistance was defined by driving signal shape change – driving signal waveform would become sinus-flank if slew-rate values fell below 1 V/μs. Current consumption in full drive operating mode (**Figure 24**) remains practically a constant value of 125 mA, independent of slew-rate practically to the limit, where the slew-rate drops to 1 V/μs. Average current measurements at high slew rates are hard to establish due to slow data processing of the oscilloscope. In such conditions, the SMPS duty cycle is constantly changing, therefore, any increase in current consumption measurements in full drive operating mode have to be attributed to measurement error. Such measurement errors could be mitigated using a larger output capacitor (**Figure 18**, C1 and C3).

When comparing both operation modes, the majority of current consumption can be attributed to SMPS power sources. In minimum drive operation mode, current consumption is reduced to 70 mA. Such minimum drive operation mode consequentially enables power-saving features during operation with reduced flow. On the other hand, full drive operation mode with its high slew rate improves the micropump air pumping capability, although, for efficient air pumping, driving frequency remains the main limitation factor.

Next, we performed the same set of measurements on a micropump, which comprises a smaller piezoelectric disc with a diameter of 6 mm and a capacitance of 4 nF [15]. Compared to the previously described 12 nF micropump, the clamping voltage (±125 V) regulation area extended from 150 Hz to 400 Hz. Both slew rates in full drive operation mode achieved levels of 22 V/μs, compared to previously achieved 16 V/μs. As expected, the current consumption in both modes reduced marginally – compared to values from **Figure 24**, —175 mA in full drive operation mode (previously 180 mA) and 160 mA in minimum drive operation mode (previously 170 mA).

3.3 Fluidic characterization

After the micropump driver testing, the system for computer-controlled characterization of piezoelectric micropumps was set up. Analyzed driving module was connected to tested micropumps, developed in our laboratory [7, 15] using the measurement setup, described in **Figure 10**. Presented micropump driving module was compared to previous driving modules, designed in our Laboratory.

Three distinct micropump designs (N, R, S), each with different outlet channel geometry, were compared using RC asymmetric/symmetric driver and both initial and final versions of the optocoupler based driver. Airflow rate and DI water flow rate measurements were performed at 100 Hz, while the presented driver was configured in full-drive operation mode in both initial and optimized versions.

Achieved air flow rate with a symmetric RC driver was 1.6 sccm. This value increased to 4.2 sccm with optocoupler based driver in full-drive operation mode. Compared with symmetric and asymmetric versions of RC driver, presented driver surpasses all previous performances. Achieved DI water flow rate with a symmetric RC waveform-like driver was 2.2 sccm. This value was increased to 2.6 sccm with optocoupler based driver in full-drive operation mode. Achieved air backpressure performance almost doubled with the use of optocoupler based driver: It is achieving its peak value of 39 mbar on N1 type micropump in 'full drive' mode. Achieved DI water backpressure performance improved by 30%, compared to initial version of RC waveform-like driver: It is achieving its peak value of 240 mbar on N1 type micropump in "full drive" mode. Both presented versions improved DI water backpressure performance over RC waveform-like drivers.

Optocoupler based driver was set up at 100 Hz with ±105 V amplitudes in full drive operation mode a micropump S29R1 was connected to its output. Slew-rate values were altered in the range from 0.2 V/μs to 16 V/μs, while both DI water and airflow rates were measured. In both cases, the flow rate is practically independent of the slew-rate (**Figure 26**).

Figure 28 is showing a slight decrease in backpressure characteristics with decreasing slew rate SR^+. This performance deteriorates severely when the slew rate falls below 4 V/μs. If slew rate is lowered even further down to 1 V/μs, comparable conditions as in the case of RC waveform-like driver may be achieved with reduced current consumption to 100 mA (see **Figure 27**) at 5 V module power supply. Even with slew rate kept as low as 1 V/μs, resulting flow rate is significantly higher

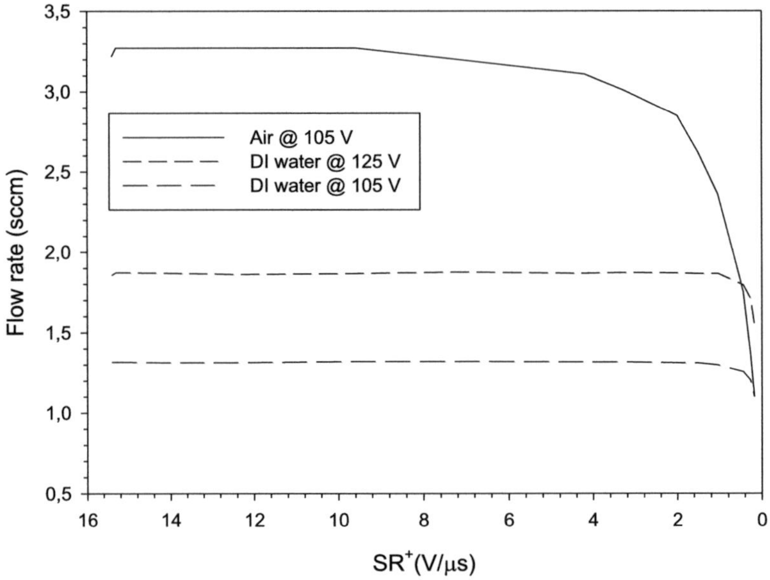

Figure 26.
Air, DI-water flow rate vs. slew rate in "full drive" mode.

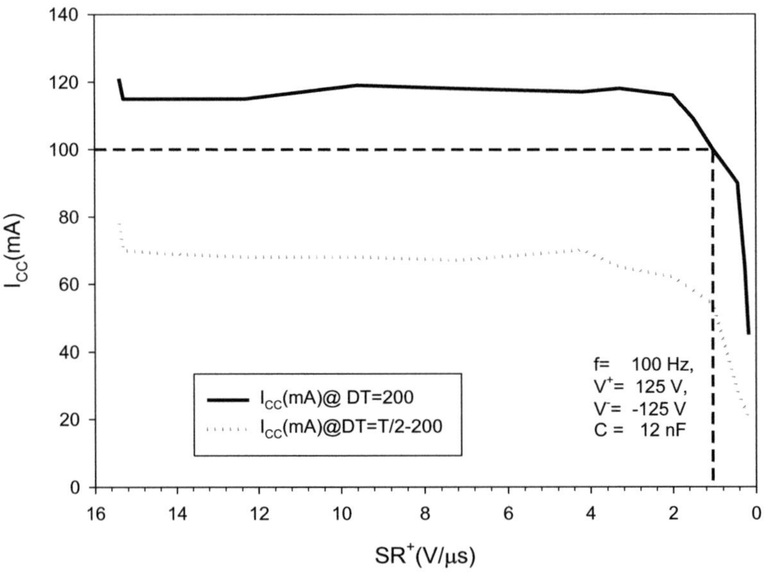

Figure 27.
Current consumption vs. positive slew-rate.

compared to our RC driving module versions, which achieved slew rates of 0.2 V/μs. Current consumption can be further reduced using minimum drive operation mode.

Starting values of SR^+ in **Figure 28** indicate that high slew rates at 16 V/μs have only a minor impact on flow rate and backpressure characteristics at a significantly higher current consumption. However, higher slew rates improve/stabilize other micropump properties and such as self-priming and bubble tolerance and enable a more reliable micropump operation in different operating conditions.

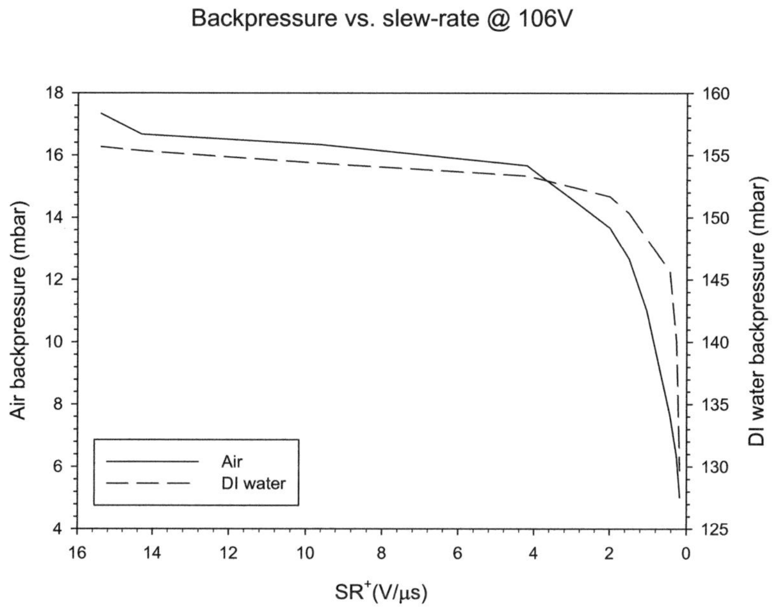

Figure 28.
Air, DI-water backpressure vs. slew-rate in "full drive" mode.

4. Miniaturized arbitrary-waveform generator design

Advanced piezoelectric micropump driver, designed as a miniaturized arbitrary waveform generator for driving custom made piezoelectric micropumps represents a large leap from initial capacitive charging/discharging square-wave shape: Developed driver enables independent setting of several output signal parameters, such as frequency, both positive/negative amplitude, both slew-rates, dead time, and modes of operation (pump/valve). Compared to optocoupler based design, dead time and slew rates are made a part of synthesized waveform and are therefore software settable in far more refined form than using resistor based D/A converter.

Air pumping characteristics of micropumps, manufactured in our laboratory, exhibited no distinct frequency peak in backpressure and flow performance. This result led to the development of a different driver, which would elevate driving frequency in order of several kHz. While the square-wave signal offered best flow and backpressure results, it also induces excessive stress on piezoelectric actuator, which might reduce its lifetime, therefore the need for other shapes of driving waveforms was established. Driving of micropumps with other signal types (e.g. sinewave), results in reduced stress on micropump piezoelectric actuator and also reduces current consumption at higher driving frequencies. Based on the previous designs, we concluded that the arbitrary waveform driver output stage should feature separate high-voltage boost stages, which are able to independently set micropump positive and negative supply voltage. Furthermore, in order to efficiently boost arbitrary micropump driving signals, a full amplification stage, based on an operational amplifier, similar to [17, 18] had to be designed instead of previously proposed interchanging boosting stages or simplified optocoupler-based square wave stage. Generating a micropump driving voltage using PWM approach has proven too slow for synthesis of driving signals above 400 Hz, therefore development in direction of D class amplifier was abandoned. Designed piezoelectric driving module would have to be capable of synthesis of arbitrary signals from DC to several kHz if we want to cover pumping of both liquids and gases. It should apply 32 kHz PWM frequency on two separate high-voltage power supplies to achieve amplitudes up to 150 V_{PP} using 5 V power supply voltage. The module parameters such as micropump excitation voltage, frequency and symmetry should be programmed using a PC. Module itself is USB powered and is to be designed with size-sensitive applications aspect. In addition to microfluidic applications, the module versatility can be extended to the evaluation of arbitrary piezo actuators in microsystems or advanced haptics.

4.1 Module operation

The concept of a separate, independently settable high voltage power supply section was already successfully implemented with the previously presented optocoupler based driver. During current consumption measurements, it was demonstrated, that the reduction of voltage step-up ratio significantly improves current driving capabilities of output stage. This design uses the same topology, capable of boosting the micropump power supply voltage in a range from ±10 V to ±150 V, depending on duty cycle setting (**Figure 18**, node label V^+ and V^-). Both SMPS converters are driven using PWM signal (**Figure 18**, sources V1 and V2) with PWM frequency 32 kHz in range from 10–80%. Additional low-voltage boost SMPS, which generates +10 V from a single +5 V power supply was added to reduce the voltage step-up ratio. Independent setting of power supply voltage levels allows adjustment of amplitude symmetry by separately adjusting the signal waveform and signal DC offset. Both high-voltage power supply outputs (**Figure 18**, V^+ and V^-) are connected to the operational amplifier high voltage power supply inputs (**Figure 29**, V^+ and V^-). High voltage boost stage comprises a single operational amplifier (U1), which feeds two common base transistor amplifiers. Positive output voltage common base amplifier (Q4 and R3) is used to translate low-output operational amplifier voltage (VAMPOUT) through R1 current onto base resistor R3, which opens positive voltage output driver Q2. Similarly, for negative output voltages, the common base amplifier (Q3 and R2) opens negative output voltage driver Q1. Note, that operation of this stage is inverting i.e. a negative swing of potential VAMPOUT causes a positive swing of output voltage VOUT. Operation of high voltage boost stage is controlled using a negative feedback loop, comprised of resistors RF1, RF2 and a capacitor C3 for frequency compensation.

Due to inverting nature of output stage, a *non-inverting input* of operational amplifier is used to complete the loop. Such feedback implements an overall non-inverting high voltage amplifier with amplification of $A_U = 42.6$, which is able to amplify input amplitude of VAMPIN = ±3.3 V to output voltage range VOUT = ±140 V.

Selection of C3 at 470 pF limits input signal (VAMPIN) bandwidth to approximately 10 kHz, a decade above frequency limits, achieved with previous versions of our driver (max. 900 Hz). Minor drawback of presented non-inverting high voltage amplifier stage is the polarity of input signal VAMPIN: In order to achieve full output voltage swing, input signal also needs to be bipolar, which is inappropriate for driving with a microcontroller fed from a single 3.3 V power supply.

To be able to use unipolar input driving signals, a separate transformation stage, depicted in **Figure 30**, was designed. Operational amplifier U2 forms a voltage subtractor, which produces an output voltage (VAMPIN) as an amplified difference

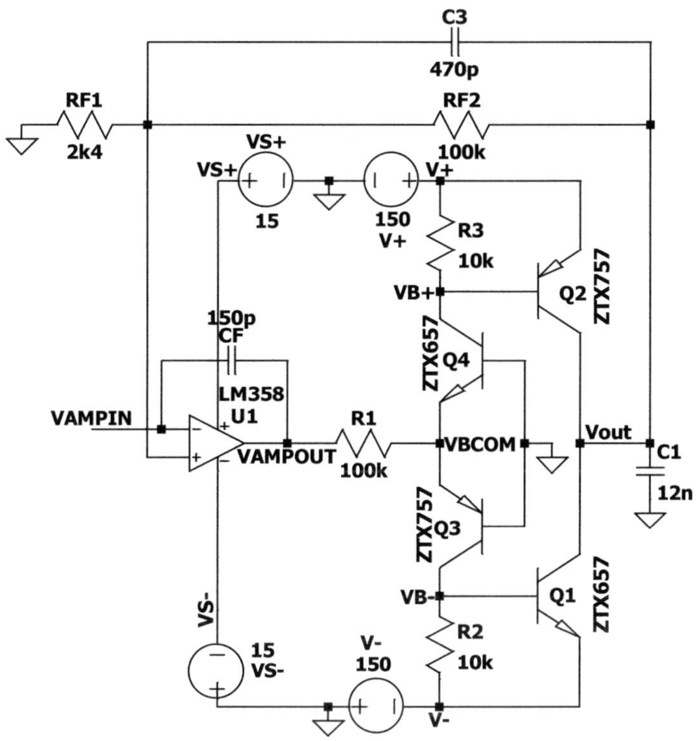

Figure 29.
Non-inverting high voltage amplifier stage.

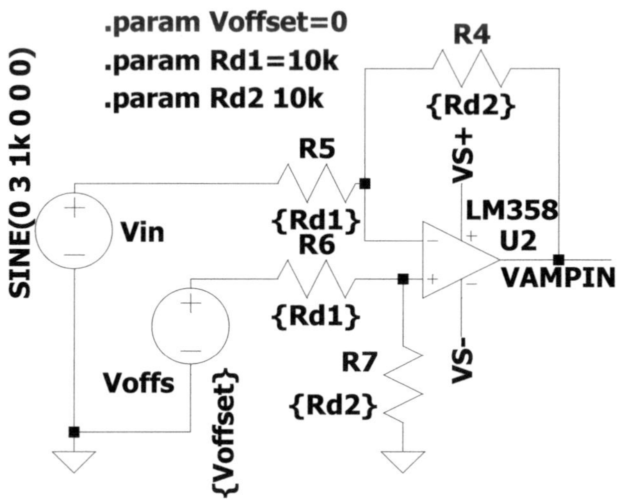

Figure 30.
Signal conditioning part of the driver.

between signal input (Vin) and offset input (Voffs). Current selection of feedback resistors R5 ... R7 values Rd1, Rd2 ratio results in amplification of $A_U = 1$.

Micropump control is achieved using a cost-effective (5€/unit) 32-bit ARM Cortex 4 microcontroller, an STM32G431KB in LQFP-48 package [19]. Selected microcontroller also features 16-bit timer (TMR2), which in combination with 12-bit buffered digital-to-analog converter (DAC1) and a 12 channel direct-memory-access (DMA) controller forms a direct digital synthesis system (DDS), capable of synthesizing digital signals with update rate of 1 MSPS. Important fact is that the DDS entirely bypasses the microcontroller core, thus saving time for additional tasks, such as communication and PWM regulation for operational amplifier power supply (**Figure 18**, VS1, VS2) and high-voltage supply voltage (V^+ and V^-) channel PWM. Selected microcontroller also features two separate 16-bit analog-to-digital converter units (ADC).

Each PWM controller uses a dedicated ADC channel in combination with one of PWM channels, available in timer (TMR1). ADC interrupt is triggered whenever conversion is complete on all four selected channels, afterwards PWM is adjusted according to power supply set-point. Microcontroller TMR1 PWM unit channels are configured independently, with four output compare registers (TMR1/CH2 and TMR1/CH3), running at 32 kHz with 12-bit resolution. PWM outputs are connected to the corresponding SMPS transistor (M1, M2 in **Figure 18**) via MOSFET driver TC4427.

Each SMPS output is monitored by feedback to the corresponding microcontroller analogue-to-digital converter input (ADC2/ADC3, **Figure 18**). After DDS generation is configured and started, it operates independently of processor actions. Input characteristic is calculated using voltage subtractor equations (see **Figure 30**) valid for ideal or DC operation, while at higher frequencies, amplification deterioration has to be taken into account. To mitigate this, the driving signal has to be monitored separately and the voltage subtractor input recalculated to actually achieved waveform positive and negative extrema. Microcontroller USB-CDC receiver was utilized as a communication interface for adjustment of all micropump driving signal parameters.

4.2 Electrical characterization

Micropump piezoelectric actuator excitation voltage was limited to ±125 V due to PZT actuator electrical field limitation and micropump frequency was set from 10 Hz to 9.2 kHz using a sinewave and square-wave excitation signal. Positive and negative driving signal amplitudes, achieved on the micropump actuator, and current consumption were measured at each micropump frequency setting. Micropump driving voltage amplitudes (V^+ and V^-) vs. excitation frequency results are presented in **Figure 31**.

In the low-frequency range, micropump driving signal clamped out to an admissible voltage limit of ±125 V, before it begins to deteriorate (**Figure 31**). In comparison to driving voltage vs. frequency scan of our previous prototypes, this clamping interval extended from 150 Hz to 1 kHz. Upper (V^+) and lower (V^-) voltage characteristics in **Figure 31** represent achievable voltage limits, which can be regulated by a voltage monitor.

Such extension of clamped-out driving signal amplitudes consequently results in an increased module current consumption I_{CC}, which is presented in **Figure 32**. Previously developed modules were capable of producing square wave driving signals in frequency range up to 700 Hz and its current consumption reached 118 mA at 100 Hz and 180 mA at 150 Hz.

Current miniaturized arbitrary waveform generator design also offers driving using sine-wave, where the true driving signal frequency extension becomes obvious: It can extend frequency driving interval up to 9.2 kHz. Current consumption using

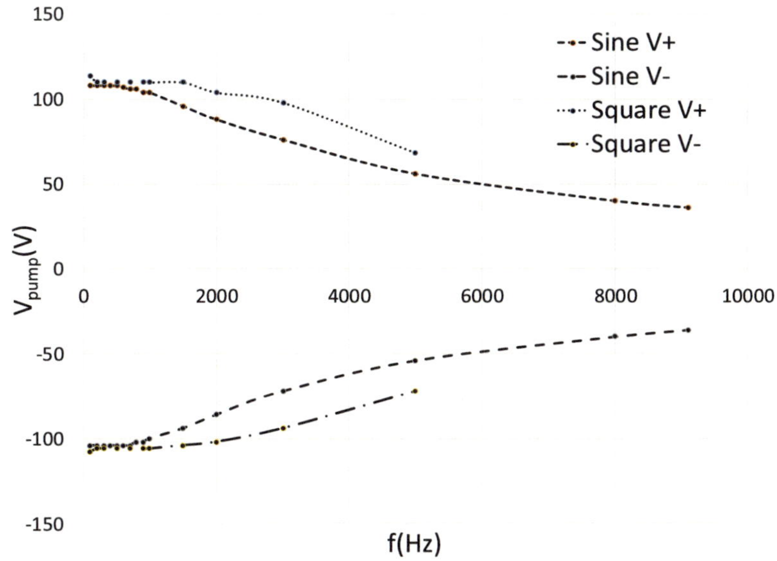

Figure 31.
Voltage -frequency sweep.

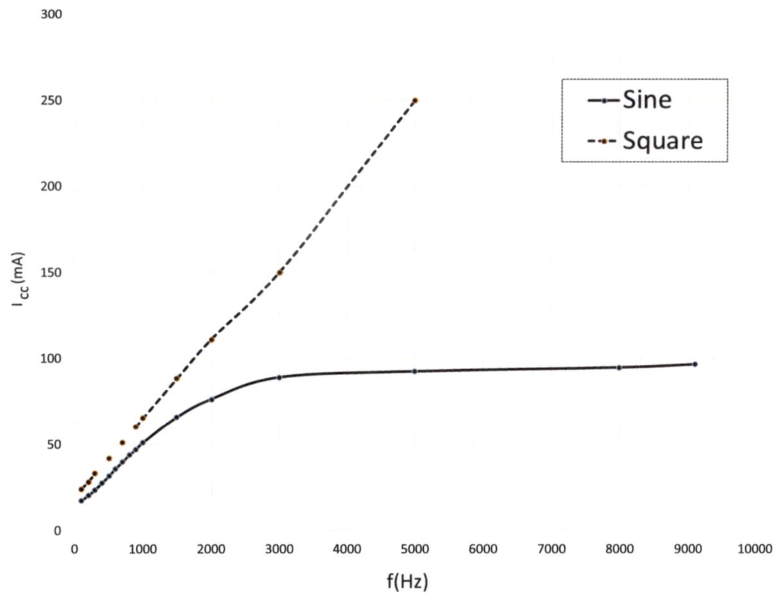

Figure 32.
Current consumption vs. frequency.

square wave driving signal is also reduced: it reaches 40 mA at 100 Hz and 75 mA at 1 kHz. A major improvement, shown in **Figure 32**, is the ability of presented driver to generate a sinewave signal, which results in a far smaller current consumption compared to square-wave driving signal. As shown in our previous research, square-wave signal offer better fluidic performance, while sinewave offer more stable micropump operation with far less stress on piezoelectric actuator.

4.3 Fluidic characterization

After initial electrical testing, fluidic characterization was performed using the same computer controlled characterization system (see **Figure 10**). Analyzed driving module was connected to tested micropumps, developed in our laboratory [7, 15]. Our previous research of micropump performance on signal shape was focused primarily on the optimization of square-wave signal parameters to improve micropump flow and backpressure performance. Impact on the slew-rate increase was demonstrated in the optocoupler based design results.

In the following measurements, we intend to focus on pumping of air, because the presented driver excelled at generating frequencies above 1 kHz, which we were unable to implement on our previous designs. Presented micropump driving module was compared to previously described driving modules, designed in our Laboratory (RC-like rectangular signal driver and optocoupler-based driver). Micropump design S32S2 with small outlet channel geometry was used for comparison between RC symmetric driver and latest version of the presented driver. Air flow rate (**Figure 33**) and DI water flow rate (**Figure 34**) measurements, were performed with sinewave signal in the frequency range from 100 Hz to 3 kHz.

Presented driver performs significantly better than previous RC waveform-like driver, especially in air pumping, where it increases the air flow rate (from 1.6 sccm to 2.2 sccm) in comparison with symmetric-amplitude RC waveform-like driver. Compared with symmetric and asymmetric versions of RC waveform-like driver, presented driver surpasses all previous performances of previously described drivers.

Compared to the optocoupler design, flow rate decrease for DI water from 3.3 ml/min to 2.2 ml/min seems substantial. Note, that 3.3 ml/min was obtained using square-wave signal with higher slew rate on optocouplers. In arbitrary waveform design, such slew rates could not be achieved not even using square-wave measurement. However, presented driver almost doubles air backpressure performance, compared to RC type driver (39 mbar) and is achieving its peak value of 52 mbar on N1 type micropump using sinewave driving signal. The flow rate and backpressure characteristic flatten out at 3 kHz, therefore this represents the usable driving frequency range for tested micropumps.

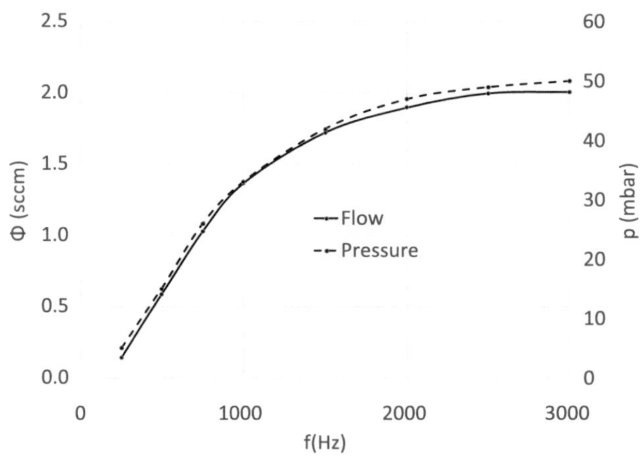

Figure 33.
Air flow rate and backpressure vs. frequency.

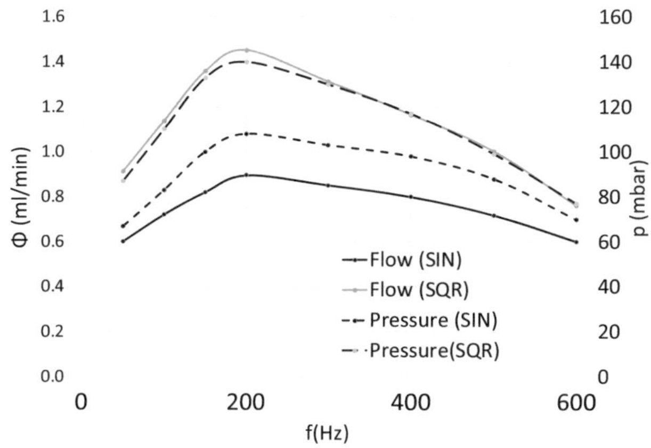

Figure 34.
DI water flow rate and backpressure vs. frequency.

5. Conclusion

Design, implementation and characterization of three distinctive piezoelectric micropump driving module designs were presented.

Our initial design was based on a miniature, transformerless version of a piezoelectric micropump driving module, based on two switched-mode power supply (SMPS) boost converters with a shared SMPS inductance and piezoelectric micropump actuator (as a common output capacitor). Its small size and its low current consumption (≈ 55 mA) make presented driver suitable for integration inside micropump housing, targeting principally cost-sensitive and low-power applications. This module synthesized driving frequencies in the range of 400 Hz while achieving amplitudes up to 250 V_{pp} (frequency range up to 80 Hz). Optimal operating frequency range for micropump actuation was found around 50 to 80 Hz during tests using DI water pumping. Optimal driving conditions considering driving module current consumption and micropump performance characteristics (power supply voltage of 9 V, excitation frequency f of 100 Hz and duty cycle DC^+ of 40%) resulted in amplitude symmetric driving signal with an amplitude of 125 V, micropump flowrate performance of 0.36 ml min^{-1}, backpressure performance of 104 mbar and module power consumption of 0.5 W. Unfortunately, this design required a separate mechanism form equalizing discrepancies in positive and negative driving amplitude.

Next, we introduced an optocoupler-based driver design, which achieved higher driving frequencies in range up to 1 kHz and amplitudes up to 250 V_{pp} (in lower frequency range up to 150 Hz), making this design optimal for pumping DI water, where positive and negative signal slew-rates up to 18 V/μs were achieved. In comparison with our previous driver design, this version eliminates the need for equalization of driving signal amplitudes. Furthermore, it increases airflow capability from 1.6 sccm to 3.3 sccm, although the air was not the primary target of pumping media for this design. Maximum module power consumption was 1.6 W (180 mA @ 9 V).

Finally, an arbitrary waveform piezoelectric micropump driver for driving custom made piezoelectric micropumps was presented. Driving signal frequency range from several Hz to 9.2 kHz was investigated and amplitudes up to 125 V_{pp} were achieved in the frequency range up to 1 kHz. Optimal micropump actuation frequency of 3 kHz for pumping air was found. Indeed, it does not achieve airflow capability of presented optocoupler based driver, however, it is capable of achieving almost double (59 mbar) backpressure at reduced current consumption of 500 mW (100 mA @ 5 V) using a sinewave driving signal. Presented modules are capable of driving a 200 μm thick piezoelectric actuator with a capacitance in span from 4 nF to 12 nF.

Acknowledgements

This research was funded by Slovenian Research Agency/ARRS, grant number P2-0244. The authors would like to thank the Slovenian Research Agency and Ministry of Education, Science and Sport for their support of this work.

Conflict of interest

The authors declare no conflict of interest.

Author details

Matej Možek[1*], Borut Pečar[1], Drago Resnik[1,2] and Danilo Vrtačnik[1]

1 Laboratory of Microsensor Structures and Electronics (LMSE), Faculty of Electrical Engineering, University of Ljubljana, Ljubljana, Slovenia

2 Chair of Microprocess Engineering and Technology–COMPETE, University of Ljubljana, Ljubljana, Slovenia

*Address all correspondence to: matej.mozek@fe.uni-lj.si

References

[1] Bartels mp-x Controller. Available from: https://www.bartels-mikrotechnik.de/elektronik/ [Accessed: November 15, 2021]

[2] Bartels mp6-OEM Controller. Available from: https://www.bartels-mikrotechnik.de/wp-content/uploads/simple-file-list/EN/Manuals-and-Data-Sheets/mp6_electronics_Datasheet.pdf [Accessed: November 15, 2021]

[3] Antonijevič M, Dolžan T, Pečar B, Aljančič U, Možek M, Vrtačnik D, et al. Characterization System for Piezoelectric Micropumps, 48th International Conference on Microelectronics, Devices and Materials & the Workshop on Ceramic Microsystems, September 19–September 21, 2012, Otočec, Slovenia. Proceedings. Ljubljana: MIDEM - Society for Microelectronics, Electronic Components and Materials. pp. 357-362

[4] Smole D, Dolžan T, Pečar B, Aljančič U, Možek M, Vrtačnik D, et al. Low-Power Piezoelectric Micropump Driving Module. 50th International Conference on Microelectronics, Devices and Materials, October 8–October 10, 2014. Ljubljana, Slovenia. pp. 139-144

[5] Baumann T, Hoene E. Highly Miniaturized Piezoactor Drive, Fraunhofer-Institut IZM. Available from: http://www.powerguru.org/highly-miniaturized-piezoactor-drive/ [Accessed: November 15, 2019]

[6] LTSPICE4. Linear Technologies. Available from: http://ltspice.linear-tech.com/software/LTspiceXVII.exe [Accessed: November 15, 2021]

[7] Dolžan T, Pečar B, Možek M, Resnik D, Vrtačnik D. Self-priming bubble tolerant microcylinder pump. Sensors and Actuators A: Physical. 2015;**233**:548-556. DOI: 10.1016/j.sna.2015.07.015

[8] 8-bit AVR Microcontroller: ATtiny4 / ATtiny5 / ATtiny9 / ATtiny10 DATASHEET. Available from: http://ww1.microchip.com/downloads/en/DeviceDoc/ATtiny4-5-9-10-Data-Sheet-DS40002060A.pdf [Accessed: November 15, 2021]

[9] Woias P. Micropumps—Past, progress and future prospects. Sensors and Actuators B: Chemical. 2005;**105**(1):28-38. DOI: 10.1016/j.snb.2004.02.033

[10] Jeong OC, Konishi S. Fabrication of all PDMS micro pump. In: IEEE International Symposium on Micro-NanoMechatronics and Human Science. IEEE; 2005. pp. 139-143. DOI: 10.1109/MHS.2005.1589977

[11] Guan Y, Ren J, Zhang G, Cheng Z. Fabrication and experiment studies of the piezoelectric micropump with saw-tooth microchannel. In: IEEE International Conference on Intelligent Computing and Intelligent Systems (Vol. 2). IEEE; 2009. pp. 733-737. DOI: 10.1109/ICICISYS.2009.5358282

[12] Guan YF, Shen MG, Han LL. Simulations and experiment analysis of a piezoelectric micropump. In: Applied Mechanics and Materials (Vol. 229). Trans Tech Publications; 2012. pp. 1688-1692

[13] Piezoelectric actuator P-5H datasheet Sunnytec Suzhou Electronics Co., Ltd. Available from: http://sunnytec-piezo.com/en/product.asp [Accessed: November 15, 2021]

[14] Pečar B, Možek M, Vrtačnik D. Thermoplastic-PDMS polymer covalent bonding for microfluidic applications. Informacije MIDEM. 2017;**47**(3):147-154

[15] Pečar B, Resnik D, Aljančič U, Vrtačnik D, Možek M, Amon S. Bubble tolerant valveless piezoelectric micropump for liquid and gasses. In: 47th International Conference on Microelectronics, Devices and Materials and the Workshop on Organic Semiconductors, Technologies and Devices, September 28–September 30, 2011, Ajdovščina, Slovenia. Proceedings. Ljubljana: MIDEM - Society for Microelectronics, Electronic Components and Materials; 2011. pp. 57-61

[16] 8-bit AVR Microcontroller ATtiny102 / ATtiny104 Datasheet complete. Available from: http://ww1.microchip.com/downloads/en/devicedoc/atmel-42505-8-bit-avr-microcontrollers-attiny102-attiny104_datasheet.pdf [Accessed: November 15, 2021]

[17] Texas Instruments AN-272 Op Amp Booster Designs. 1981. Available from: https://www.ti.com/lit/an/snoa600b/snoa600b.pdf [Accessed: November 15, 2021]

[18] Analog Devices, Power Gain Stages for Monolithic Amplifiers. 1986. Available from: https://www.analog.com/media/en/technical-documentation/application-notes/an18f.pdf [Accessed: November 15, 2021]

[19] ST Microelectronics Devices, Mainstream Arm Cortex-M4 MCU 170 MHz with 128 Kbytes of Flash memory, Math Accelerator, Medium Analog level integration. Available from: https://www.st.com/en/microcontrollers-microprocessors/stm32g431kb.html [Accessed: November 15, 2021]

Section 4

Modeling and Control

Chapter 6

Topology Optimization Methods for Flexure Hinge Type Piezoelectric Actuators

Shitong Yang, Yuelong Li, Guangda Qiao, Xiaosong Zhang and Xiaohui Lu

Abstract

Piezoelectric actuators have the obvious advantages of simple and compact structure, high precision and long stroke. However, it is difficult to satisfy the various industrial requirements. Topology optimization method can be used to find the new configurations of the compliant mechanism, and different objective function and constraint conditions can be flexibly used to determine the compliant mechanism. In the research of piezoelectric actuators, due to the advantages of compact structure, no lubrication and large displacement magnification, compliant mechanism is extremely suitable to be introduced into the design of piezoelectric actuators. In recent years, topology optimization method is frequently used to design the compliant mechanism on piezoelectric actuator, and has become a research hotspot. In this chapter, the development of topology optimization method is introduced, the design and research on the compliant mechanism of piezoelectric actuator have been summarized, and the future research direction and challenges of topology optimization design for flexure hinge type piezoelectric actuators are prospected.

Keywords: topology optimization, piezoelectric actuator, flexure hinge, compliant mechanism, displacement amplification

1. Introduction

Topology optimization method is firstly developed by Bendsøe and Kikuchi [1] and Bendsøe [2], which is originally used for an elastic material distribution problem in 1988. Compared with sizing and shape optimization, topology optimization is a more powerful tool, which allows new holes and connections to be generated in the design domain with a prescribed amount of material. Recently, the topology optimization method has been extended to many other fields, such as materials science [3, 4], micro-machines [5, 6], precision and ultra-precision machining [7, 8], optical focusing [9, 10], and so on. With the development of computer technology, topology optimization is developing very fast, and the calculation process of topology optimization problem has been significantly simplified.

In the research of piezoelectric actuators, piezoelectric stacks are usually used in the driving field, which requires precise but minute motion [11, 12]. Flexure hinge type piezoelectric actuators are widely applied in precision machining, in-situ mechanical measurement, biomedicine and other fields because of its compact structure, long stroke and fast response speed [13–15]. The piezoelectric actuators are usually composed of piezoelectric stack, compliant mechanism and slider [16–20], here the compliant mechanisms are used as transmission and amplifying mechanisms. In the field of piezoelectric actuators, the modeling and design of compliant mechanisms are key issues. For example, bridge-type [21] and parallelogram-type [22] compliant mechanisms are applied to the reach of precision engineering. Sigmund [23, 24] has designed an inverting displacement amplifier, which has been used to obtain the maximum mechanical advantage. In addition, topology optimization methods have been assembled in commercial software [25], which makes the applications of topology optimization methods much easier.

There mainly three different types of topology optimization method are used to handle the design objectives and constraints of piezoelectric actuators: density-based methods, boundary variation methods, hard-kill methods, respectively [25, 26]. (1) Density-based methods, which include Solid Isotropic Material with Penalization (SIMP) and Rational Approximation of Materials Properties (RAMP) technique. (2) Boundary variation methods (level set and phase field). (3) Hard-kill methods, typically Evolutionary Structural Optimization (ESO) method and Bidirectional Evolutionary Structural Optimization (BESO) method. Many scholars have conducted in-depth research on the above three types of topology optimization method, the design theories and methods have developed rapidly. Topology optimization has becoming one of the important methods of piezoelectric actuator design.

However, topology optimization of piezoelectric actuators is a complicated problem, which contains the process of determining the connectivity, shape, and location of voids inside a given design domain. Traditional researches have given some topological structure for the design of piezoelectric actuators [27–34], the not solved problem is that which kind of compliant mechanisms with the flexure hinges makes the output performances best. Yang et al. [35] developed a static topology optimization method to solve the problem, and some topology optimization methods have also been extended to design the compliant mechanism of piezoelectric actuators [36–40]. At present, there is little research on using topology optimization method to design flexible hinges, and its theories and methods are very scarce, which has a lot of exploration space. Therefore, it is particularly important to explore the theories and methods of topology optimization for the design of flexure hinge type piezoelectric actuator.

We hope that this chapter will provide a summary of the recent advances and novel applications of topology optimization methods, provide exposure to lesser known, yet promising, techniques, and serve as a resource for those new to the field. The presentation of each method focuses on new developments and novel applications. So in Section 2, the operating principle of piezoelectric actuators is introduced, the problems are described and the topology optimization model is established, then the simulation analysis is performed. The prototype is fabricated and the experimental system is built in Section 3. Systematic experimental test is conducted to study the actual performance of the actuator in Section 4, and the conclusion and discussion of this chapter are in Section 5.

2. Operating principle and topology optimization method

2.1 Operation principle

The operating principle of the piezoelectric stick–slip actuator designed by the topology optimization method is shown in **Figure 1**. The asymmetric sawtooth wave is used as the excitation signal. A motion cycle can be divided into three phases:

Initial phase (as shown in **Figure 1a**): At time t_0, the input voltage is zero, so the piezoelectric stack is at its original length, the actuator is at rest, and the maximum static friction force between the indenter and slider is taken as the locking force F_0, point P is the contact point between the indenter and slider.

Stick phase (as shown in **Figure 1b**): From time t_0 to t_1, the length of piezoelectric stack extends with the increases of voltage. Under the effect of flexure hinge, the driving mechanism produces oblique deformation. The vertical positive pressure Fy_1

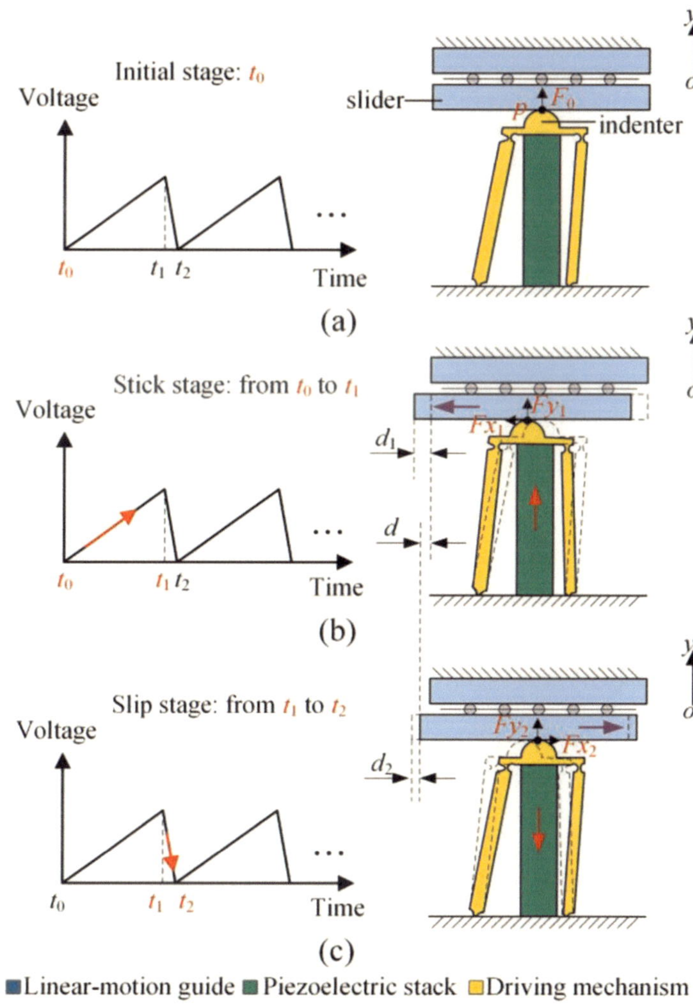

Figure 1.
Operating principle: (a) initial phase. (b) Stick phase. (c) Slip phase.

on the slider increases, and the horizontal driving force Fx_1 drives the slider to move d_1 distance by static friction force, resulting in a large output displacement. When it reaches time t_1, the elongation of piezoelectric stack reaches the maximum.

Slip phase (as shown in **Figure 1c**): Contrary to the stick stage, the piezoelectric stack shrinks rapidly with the sudden drop of the input voltage, the driving mechanism returns to its original shape, and the vertical positive force Fy_2 of the indenter against the slider decreases. There is relative sliding between the stator and the slider, resulting in horizontal kinetic friction force Fx_2 in the opposite direction. When the action of kinetic friction force is greater than that of inertial force, a small backward displacement d_2 occurs.

After one motion cycle, the slider moves a net forward distance d, and the actuator realizes a large-stroke output by repeating the above process driven by periodic asymmetric sawtooth wave.

2.2 Problem description

In order to better describe the method, the symbols used in the structural topology optimization process are expressed in **Table 1**.

Bendsøe and Sigmund pointed out that the ratio between output and input displacements is an important objective function for compliant mechanism [25]. The problem is how to find the optimal driving mechanism shape that can achieve the design goal in a certain design domain, and we design the driving mechanism on the basis of the theories of Ref. [25, 40].

Name	Symbol	Name	Symbol
Displacement vector	$\mathbf{u}$	Design domain volume	V_0
Input displacement	u_{in}	Prescribed volume fraction	V_f
Output displacement	u_{out}^x	Volume of element	v_e
Parasitic displacement	u_{out}^y	Total number of elements	N
Global stiffness matrix	$\mathbf{K}$	Penalization power	p
Element stiffness matrix	$\mathbf{k}_e$	Displacement magnification factor	λ
The element stiffness matrix for an element with unit Young's modulus	$\mathbf{k}_0$	Evaluation indicator	τ
Global element stiffness matrix	$\mathbf{K}_e^0$	Minimum relative density value	$\rho_{\min}$
Input stiffness matrix	$\mathbf{K}_{in}$	Objective function	g
Output stiffness matrix	$\mathbf{K}_{out}$	Constraint function	g_τ, g_v
Input stiffness	k_{in}	Adjoint unit load vector	$\mathbf{L}_x$, $\mathbf{L}_y$, $\mathbf{L}_{in}$
Output stiffness	k_{out}	Adjoint displacement vector	$\overline{\mathbf{u}}_x$, $\overline{\mathbf{u}}_y$, $\overline{\mathbf{u}}_{in}$
Input force vector	$\mathbf{f}_{in}$	Element of adjoint displacement vector	$(\overline{\mathbf{u}}_x)_e$, $(\overline{\mathbf{u}}_y)_e$, $(\overline{\mathbf{u}}_{in})_e$
Input force	f_{in}	Element of displacement vector	$\mathbf{u}_e$

Table 1.
The symbols of topology optimization method.

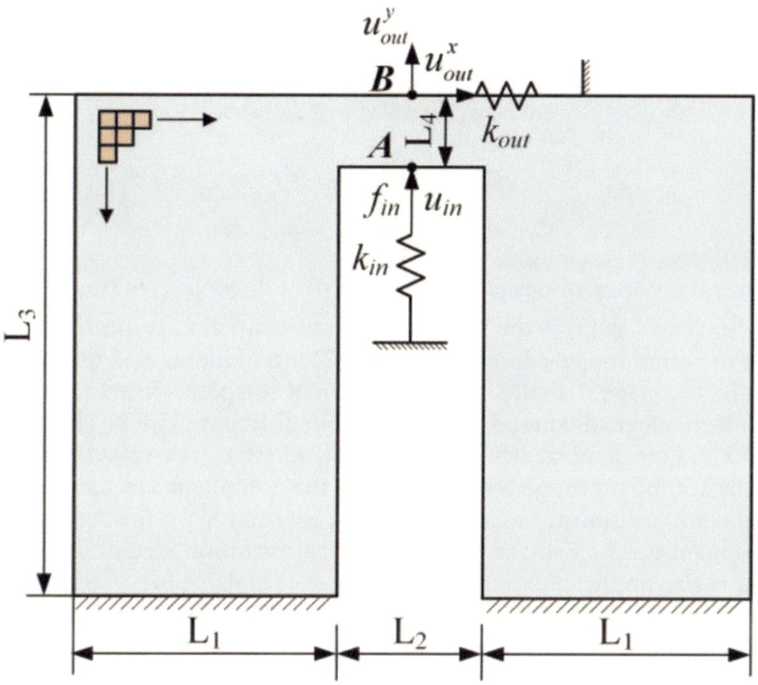

Figure 2.
The design domain of driving mechanism.

The structural characteristics of the stick–slip actuator are considered, the design domain of the flexure hinge mechanism is revealed in **Figure 2**, and the dimensions of the design domain are L_1 = 12.5 mm, L_2 = 7 mm, L_3 = 24 mm, L_4 = 3.5 mm, respectively. u_{out}^{x}, u_{out}^{y}, and u_{in} are the output displacement at point B in x direction, y direction and the input displacement at point A, respectively. f_{in} is the input force, k_{in} and k_{out} are the stiffness of input spring and output spring, respectively.

The magnification factor λ is defined as the objective function, which is considered the relationship of output displacement u_{out}^{x} and input displacement u_{in}, while the parasitic displacement u_{out}^{y} is small, here the evaluation indicator τ is defined as a constraint condition to restrict the parasitic displacement u_{out}^{y}.

The displacement magnification factor λ and evaluation indicator τ are given as follows:

$$\lambda = \frac{u_{out}^{x}}{u_{in}} \tag{1}$$

$$\tau = \frac{u_{out}^{y}}{u_{out}^{x}} \tag{2}$$

2.3 Mathematical formulation

A basic engineering goal of piezoelectric stick–slip actuators is to maximize the stroke amplification in structures and mechanisms, a serious of different objective functions can be formulated. The basic topology optimization model of piezoelectric actuator can be given as:

$$\begin{aligned}
&\text{find}\ \ \boldsymbol{\rho} = [\rho_1, \rho_2, \cdots, \rho_N]^T \in R^n \\
&\text{min}\ \ g(\boldsymbol{\rho}) \\
&\text{s.t.}\ \ \left\{g_j(\boldsymbol{\rho}) \leq 0, j = 1, 2, \cdots, m\right\} \\
&\qquad g_v(\boldsymbol{\rho}) = \int_{\Omega} \rho dV \leq V_f V_0 \\
&\qquad \rho_i = 0\ or\ 1, i = 1, 2, \cdots, n.
\end{aligned} \tag{3}$$

where $\boldsymbol{\rho}$ is the vector of design variables, $g(\boldsymbol{\rho})$ is the objective function, $g_j(\boldsymbol{\rho})$ is the constraint functions, $g_v(\boldsymbol{\rho})$ is the volume constraint function respectively.

The optimization models described in Eq. (3) are implemented using the MATLAB programming language. Usually, the design aim of the piezoelectric actuators is to maximize mechanical advantage, so the objective functions can be chosen as magnification factor and mechanical efficiency [23, 24], which can be used for static response. Although the compliant hinge is only a part of the compliant mechanism, and its motion and configuration are relatively simple, we can regard the compliant hinge as a simple compliant mechanism, so the topology optimization idea of the compliant mechanism is also applicable to the design of the compliant hinge. Sigmund [23, 24] developed an inverting displacement amplifier, and in [25] mechanical efficiency (ME) is defined as an objective function.

The objective function of the piezoelectric stick–slip actuator topology optimization problem is to obtain largest magnification factor λ, and the evaluation indicator τ is used to restrict the parasitic displacement u_{out}^y. So the topology optimization model is given as follows:

$$\begin{aligned}
&\text{find}\ \ \boldsymbol{\rho} = [\rho_1, \rho_2, \cdots, \rho_N]^T \in R^N \\
&\text{min}\ \ g(\boldsymbol{\rho}) = -\frac{u_{out}^x}{u_{in}} \\
&\text{s.t.}\ \ \mathbf{Ku} = \mathbf{f}_{in} \\
&\qquad g_\tau(\boldsymbol{\rho}) = \left(\frac{u_{out}^y}{u_{out}^x}\right)^2 \leq \tau^* \\
&\qquad g_v(\boldsymbol{\rho}) = \sum_{e=1}^{N} v_e \rho_e \leq V_f V_0 \\
&\qquad 0 < V_f < 1 \\
&\qquad 0 < \rho_{\min} \leq \rho_e \leq 1, e = 1, 2, \cdots, N.
\end{aligned} \tag{4}$$

where $\mathbf{K}$ is the global stiffness matrix, $\mathbf{u}$ is the global displacement vector and $\mathbf{f}_{in}$ is input force vector, respectively. $\boldsymbol{\rho}$ is the design variables vector, $\rho_{\min}$ (usually $\rho_{\min} = 0.001$) is introduced to avoid singularity case. v_e is the volume of element e, V_0 and V_f are the design domain volume and the prescribed volume fraction, N is the number of elements used to discretize the design domain, respectively.

The element stiffness matrix $\mathbf{k}_e$ can be written as

$$\mathbf{k}_e = \rho_e^p \mathbf{k}_0 \tag{5}$$

where p represents the penalization power (typically p =3), and $\mathbf{k}_0$ indicates the element stiffness matrix for an element with unit Young's modulus.

The global stiffness matrix $\mathbf{K}$ can be expressed as

$$\mathbf{K} = \sum_{e=1}^{N} \rho_e^p \mathbf{K}_e^0 + \mathbf{K}_{in} + \mathbf{K}_{out} \tag{6}$$

where $\mathbf{K}_e^0$ is the element stiffness matrix for an element with unit Young's modulus in global sense, $\mathbf{K}_{in}$ is the stiffness matrix of the input spring k_{in} at the global level, $\mathbf{K}_{out}$ is the stiffness matrix of the output spring k_{out} at the global level, respectively.

2.4 Sensitivity analysis

In this part, how to determinate the output displacement u_{out}^x, parasitic displacement u_{out}^y and input displacement u_{in} and their derivatives of $\frac{\partial u_{out}^x}{\partial \rho_e}$, $\frac{\partial u_{out}^y}{\partial \rho_e}$ and $\frac{\partial u_{in}}{\partial \rho_e}$, is a key problem. In addition, the relationships of u_{out}^x, u_{out}^y and u_{in} can be considered as.

$$u_{out}^x = \mathbf{L}_x{}^T \mathbf{u},\ u_{out}^y = \mathbf{L}_y{}^T \mathbf{u},\ u_{in} = \mathbf{L}_{in}{}^T \mathbf{u} \tag{7}$$

where $\mathbf{L}_x$, $\mathbf{L}_y$ and $\mathbf{L}_{in}$ are the vectors for which the inner product with $\mathbf{u}$ produces the relevant output displacement u_{out}^x, u_{out}^y and the relevant input displacement u_{in} ($\mathbf{L}_x$, $\mathbf{L}_y$ and $\mathbf{L}_{in}$ are interpreted as a (unit) load vector), respectively.

The derivatives of u_{out}^x, u_{out}^y and u_{in} can be expressed as

$$\frac{\partial u_{out}^x}{\partial \rho_e} = \mathbf{L}_x{}^T \frac{\partial \mathbf{u}}{\partial \rho_e} \tag{8}$$

$$\frac{\partial u_{out}^y}{\partial \rho_e} = \mathbf{L}_y{}^T \frac{\partial \mathbf{u}}{\partial \rho_e} \tag{9}$$

$$\frac{\partial u_{in}}{\partial \rho_e} = \mathbf{L}_{in}{}^T \frac{\partial \mathbf{u}}{\partial \rho_e} \tag{10}$$

Differentiating the static equation $\mathbf{K}\mathbf{u} = \mathbf{f}_{in}$, we have

$$\frac{\partial \mathbf{u}}{\partial \rho_e} = \mathbf{K}^{-1}\left(\frac{\partial \mathbf{f}_{in}}{\partial \rho_e} - \frac{\partial \mathbf{K}}{\partial \rho_e}\mathbf{u}\right) \tag{11}$$

Due to $\mathbf{f}_{in}$ is a permanent load vector, we obtain

$$\frac{\partial \mathbf{f}_{in}}{\partial \rho_e} = 0 \tag{12}$$

Differentiating the global stiffness matrix, we have

$$\frac{\partial \mathbf{K}}{\partial \rho_e} = p\rho_e^{p-1}\mathbf{K}_e^0 = p\rho_e^{p-1}\mathbf{k}_0 \tag{13}$$

Next, we need to solve the adjoint equilibrium equations.

$$\mathbf{K}\overline{\mathbf{u}}_x = \mathbf{L}_x,\ \mathbf{K}\overline{\mathbf{u}}_y = \mathbf{L}_y, \mathbf{K}\overline{\mathbf{u}}_{in} = \mathbf{L}_{in} \tag{14}$$

where $\overline{\mathbf{u}}_x$, $\overline{\mathbf{u}}_y$ and $\overline{\mathbf{u}}_{in}$ are the adjoint displacement vectors, which are determined by adjoint (unit) load vector $\mathbf{L}_x$, $\mathbf{L}_y$ and $\mathbf{L}_{in}$, respectively.

Combined with Eq. (8), Eqs. (11)–(14), $\frac{\partial u_{out}^x}{\partial \rho_e}$ can be written as

$$\begin{aligned}\frac{\partial u_{out}^x}{\partial \rho_e} &= \mathbf{L}_x{}^T \frac{\partial \mathbf{u}}{\partial \rho_e} = \mathbf{L}_x{}^T \mathbf{K}^{-1}\left(\frac{\partial \mathbf{f}_{in}}{\partial \rho_e} - \frac{\partial \mathbf{K}}{\partial \rho_e}\mathbf{u}\right) \\ &= \left(\mathbf{K}^{-1}\mathbf{L}_x\right)^T\left(\frac{\partial \mathbf{f}_{in}}{\partial \rho_e} - \frac{\partial \mathbf{K}}{\partial \rho_e}\mathbf{u}\right) = -\overline{\mathbf{u}}_x{}^T \frac{\partial \mathbf{K}}{\partial \rho_e}\mathbf{u} \\ &= -p\rho_e^{p-1}\overline{\mathbf{u}}_x{}^T \mathbf{K}_e^0 \mathbf{u} = -p\rho_e^{p-1}(\overline{\mathbf{u}}_x)_e^T \mathbf{k}_0 \mathbf{u}_e\end{aligned} \tag{15}$$

Then we have

$$\frac{\partial u_{out}^y}{\partial \rho_e} = -p\rho_e^{p-1}\left(\overline{\mathbf{u}}_y\right)_e^T \mathbf{k}_0 \mathbf{u}_e \tag{16}$$

$$\begin{aligned}\frac{\partial u_{in}}{\partial \rho_e} &= -p\rho_e^{p-1}(\overline{\mathbf{u}}_{in})_e^T \mathbf{k}_0 \mathbf{u}_e \\ &= -p\rho_e^{p-1}\frac{\mathbf{u}_e{}^T \mathbf{k}_0 \mathbf{u}_e}{\|\mathbf{f}_{in}\|}\end{aligned} \tag{17}$$

where $(\overline{\mathbf{u}}_x)_e$, $\left(\overline{\mathbf{u}}_y\right)_e$, $(\overline{\mathbf{u}}_{in})_e$ and $\mathbf{u}_e$ are the elements of displacement vectors of $\overline{\mathbf{u}}_x$, $\overline{\mathbf{u}}_y$, $\overline{\mathbf{u}}_{in}$ and $\mathbf{u}$, respectively.

The sensitivity of objective function g is found as

$$\frac{\partial g}{\partial \rho_e} = -\frac{\frac{\partial u_{out}^x}{\partial \rho_e} u_{in} - \frac{\partial u_{in}}{\partial \rho_e} u_{out}^x}{u_{in}{}^2} \tag{18}$$

and the sensitivity of constraint functions g_τ and g_v can be expressed as

$$\frac{\partial g_\tau}{\partial \rho_e} = 2\left(\frac{u_{out}^y}{u_{out}^x}\right)\frac{\frac{\partial u_{out}^y}{\partial \rho_e} u_{out}^x - \frac{\partial u_{out}^x}{\partial \rho_e} u_{out}^y}{u_{out}^x{}^2} \tag{19}$$

$$\frac{\partial g_v}{\partial \rho_e} = v_e \tag{20}$$

To avoid the checkerboards patterns and mesh dependencies phenomena, some restriction on the design must be imposed. Here a filtering technique is used to modify the sensitivity of $\frac{\partial g}{\partial \rho_e}$ as follows:

$$\frac{\widehat{\partial g}}{\partial \rho_e} = \frac{1}{\rho_e \sum_{k=1}^{N_k} H_k^e} \sum_{k=1}^{N_k} H_k^e \rho_k \frac{\partial g}{\partial \rho_k} \tag{21}$$

where N_k is the set of elements e, $dist(e,k)$ implies the center-to-center distance of element e and element k, which is smaller than the filter radius $r_{\min}$, and the convolution operator (weight factor) H_k^e in Eq. (21) is defined as

$$H_k^e = \max\ (0, r_{\min} - dist(e,k)) \tag{22}$$

From Eq. (7) and Eqs. (15)–(17), the solutions of Eqs. (18), (19) are obtained.

Figure 3 shows the flow chart of topology optimization procedures. Initially, the design domain, boundary conditions and material are defined, the design domain is discretized into finite element meshes and solve the static problem. Then the sensitivities of the objective function and constraint conditions are obtained, design variables are updated by MMA algorithm. Here the MMA algorithm is employed by Svanberg [41, 42] as the optimizer. Finally, check the convergence of the results, if it converges, the iteration will stop; if not, go to step 3.

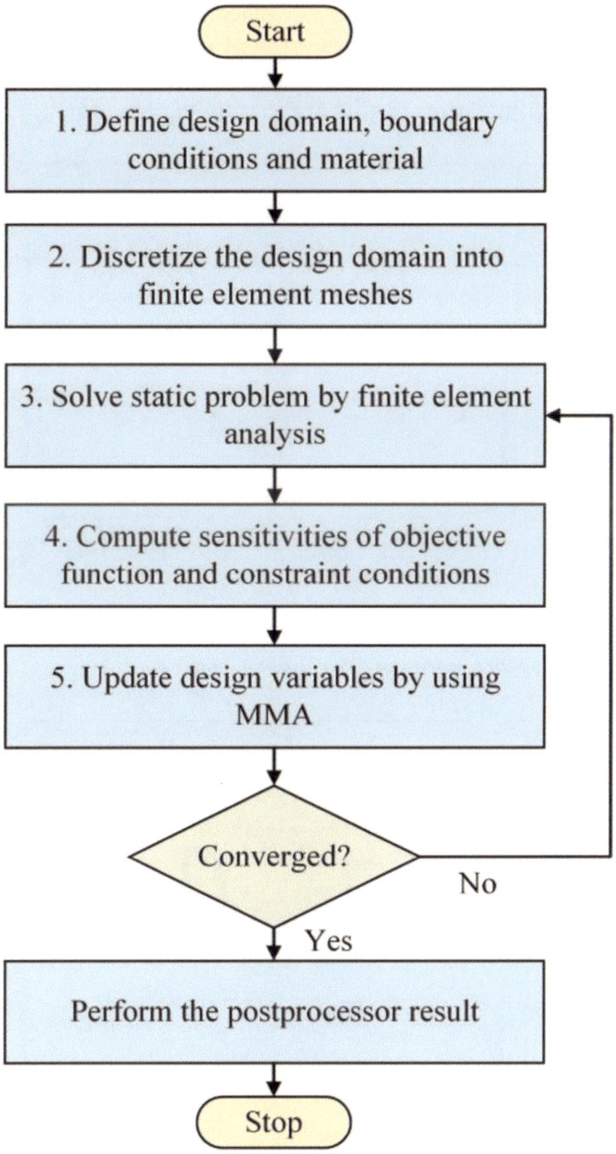

Figure 3.
Flow chart of topology optimization procedures.

2.5 Numerical results and structural design

Figure 2 shows that the design domain is divided into 64 × 48 square four node elements, the side length of the element is 1, and the structural dimensions and material units are dimensionless. Based on the practical engineering experiments, the parameters in topology optimization model are chosen as follows: the volume fraction V_f is set to 0.2, τ^* is set to 10^{-4}, and the input stiffness k_{in} and output stiffness k_{out} in Eq. (6) are chosen as 10 and 0.001, respectively. and the input force is a unit force. For more details, we refer readers to Refs. [25, 40].

Figure 4 shows that the evaluation indicator τ decrease significantly with the increase of iteration number. Therefore, the first constraint restriction in Eq. (4) markedly eliminates the effect of parasitic displacement u_{out}^{y} on the output displacement u_{out}^{x}. As shown in **Figure 5**, the displacement magnification factor λ raises with the increases of iteration number. With the continuous updating of design variables, the related topology optimization results are gradually converged.

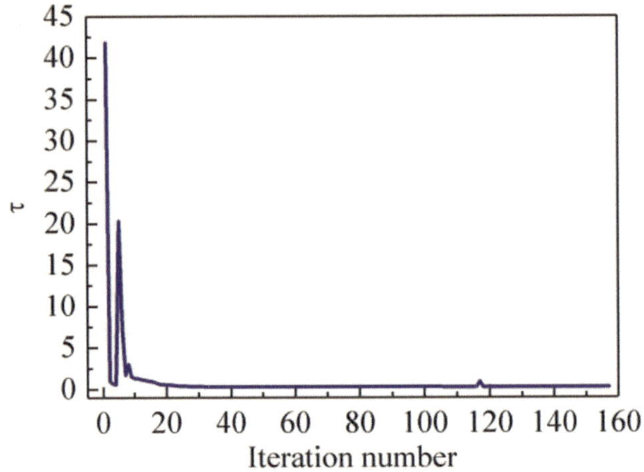

Figure 4.
The relationship between the iteration number and evaluation indicator τ.

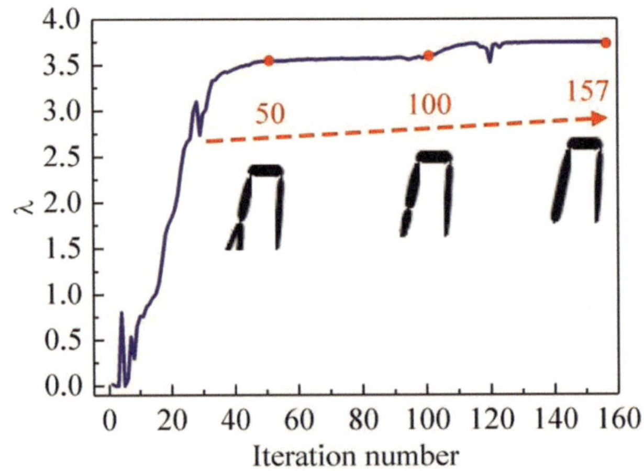

Figure 5.
The relationship between the iteration number and displacement magnification factor λ.

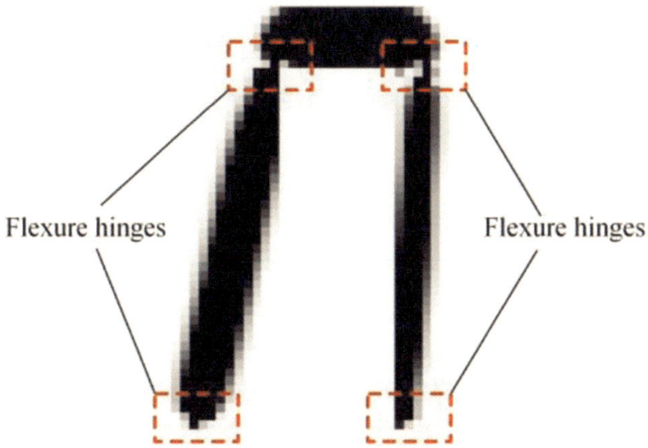

Figure 6.
The result of the topology optimization.

Figure 6 shows the topology optimization result, which is similar to the traditional four-bar mechanism, and the positions of the four flexure hinges and the tilt angles of the beams will be the main consideration in the mechanism design. Due to the gray unit in the topology image and the inaccurate contour extraction, the details of the hinges and boundaries of the structure are modified. **Figure 7** showns the main parameters of the driving mechanism, the overall dimension of the flexure hinge mechanism is 32 mm × 32 mm × 9 mm, and an indenter with radius of 2.5 mm is

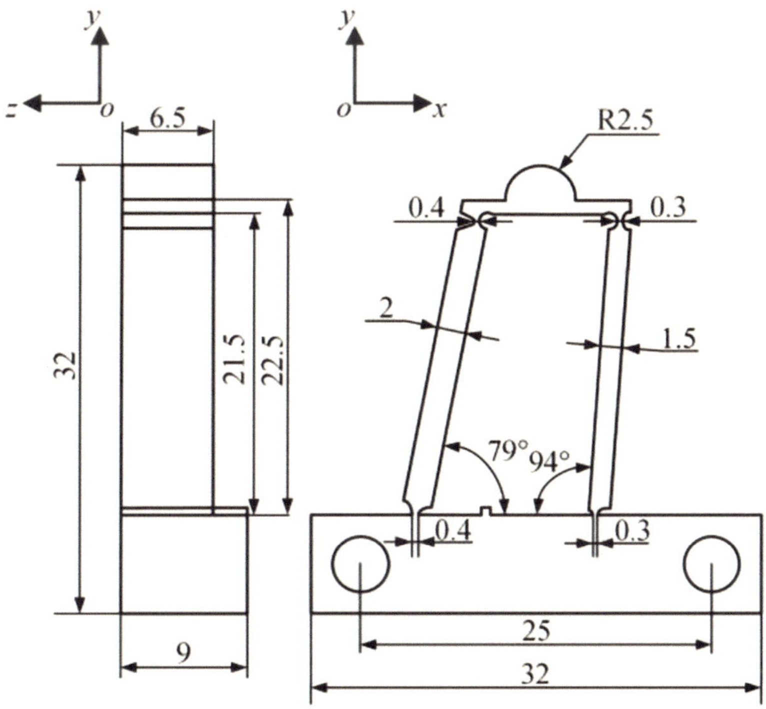

Figure 7.
Structural parameters of the driving mechanism. (unit: mm).

selected to ensure that the driving mechanism can provide sufficient contact. The thicknesses of the four special-shaped flexure hinges are 0.3 mm, 0.3 mm, 0.4 mm and 0.4 mm, respectively.

2.6 Finite element analysis

Before the experimental verification, it is necessary to ensure that the flexure hinge driving mechanism can achieve the desired displacement amplification effect, thus the finite element analysis is carried out. AL7075 is selected as the material of the driving mechanism, the Young's modulus, density and Poisson's ratio of this material are 7.17×10^4 MPa, 2810 kg/m^3 and 0.33, respectively. Take the contact point P at the top of the driving mechanism as the reference, and the two holes are defined as fixed constraints. The elongation of the piezoelectric stack is set to 10 μm to simulate the deformation of piezoelectric elements. The static simulation deformation is displayed in **Figure 8a**, the total displacement from point P to P' is 52.415 μm, the displacement in the positive direction of y-axis direction is 10.006 μm, and the displacement along the negative direction of the x-axis direction is 51.195 μm. Then the analysis of equivalent stress is shown in **Figure 8b**, the maximum equivalent stress of the driving mechanism is 128.68 MPa near the point Q, which is less than the allowable stress of AL7075 to ensure the safe operation of the actuator. The static simulation results verify the feasibility and reliability of the flexure hinge driving mechanism. Further, the first-order mode of the mechanism is shown in **Figure 8c**. To ensure the operation stability of the actuator, the driving frequency should be lower than this value as much as possible.

3. Prototype configuration and experimental system construction

The prototype configuration of the developed actuator is shown in **Figure 9**. It mainly includes a slider, a stator, a preload mechanism and a base. The stator contains a piezoelectric stack, a driving mechanism, a preload screw and a shim block. The locking force is the maximum static friction force between the indenter and the slider, it can be adjusted by the preload mechanism. One end of the linear-motion guide is a slider for precise long-stroke motion, and the other end is fixed on the base. The base is used for fixing and supporting.

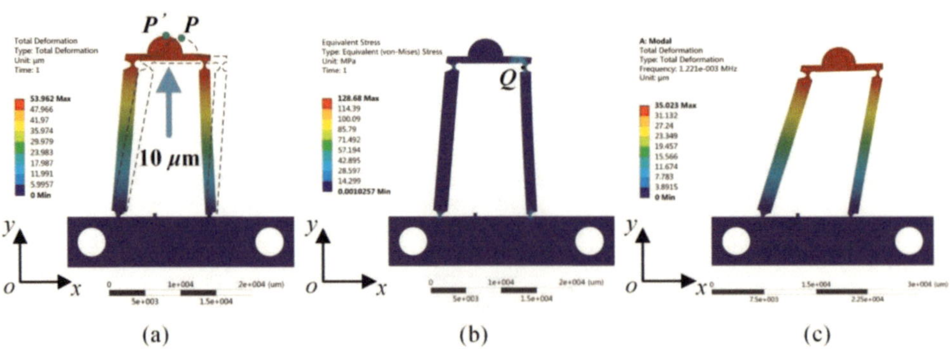

Figure 8.
Simulation analysis by FEA: (a) the static simulation deformation. (b) Analysis of equivalent stress. (c) First-order mode of the mechanism.

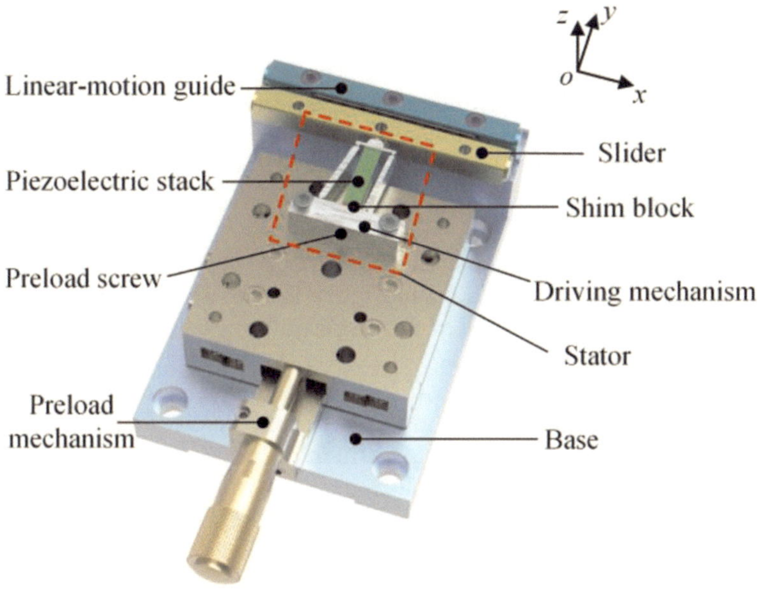

Figure 9.
Prototype configuration of the developed actuator.

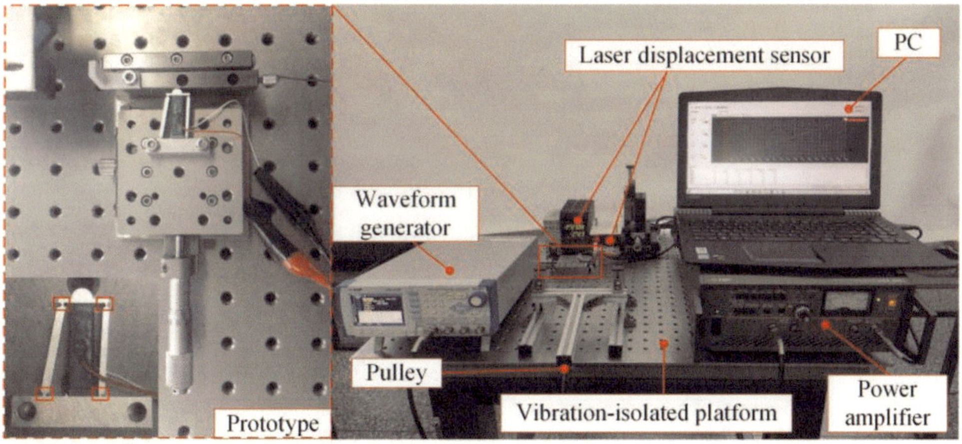

Figure 10.
The experimental system and the prototype.

As shown in **Figure 10**, an experimental system is constructed to study the motion characteristics of the actuator. The experimental system mainly includes the prototype, a waveform generator, a power amplifier, a laser displacement sensor, a PC, a pulley, and a vibration-isolation platform. The asymmetric sawtooth wave is sent out by the waveform generator (WF 1974, Negative Feedback Corporation), then adjusted by the power amplifier (HSA 4051, Negative Feedback Corporation) as the excitation signal for the piezoelectric stack (AE0505D16). The motion characteristics are detected by the laser displacement sensor (LK-H020, Keyence Corporation), and saved in the PC. The pulley and wire are used to connect the slider and the weight, and the weight is used to calibrate the locking force and apply the load.

4. Characteristic experiments

The driving voltage is set to 100 V_{p-p}, and the frequency characteristics of the actuator under different locking forces are tested and shown in **Figure 11**. Under the locking force of 1 N, the maximum velocity is 15.25 mm/s at the frequency of 650 Hz; when the locking forces are 2 N and 3 N, the maximum velocities can be obtained at the frequency values of 700 Hz and 690 Hz, and the velocities are 12.48 mm/s and 9.67 mm/s, respectively.

Thus, there are corresponding different optimal frequencies under the different locking forces conditions. As displayed in **Figure 12**, the voltage characteristics under different locking forces are explored at the corresponding optimal frequency. It can be seen that the minimum starting voltage of the actuator increases with the increases of the locking force, and the minimum starting voltage of the actuator is 21.5 V_{p-p} under the locking force of 1 N. In addition, the elongation of the piezoelectric stack increases as the voltage raises, the single step displacement of the actuator increases, thereby increasing the velocity. The motion resolution is also an important parameter of the actuator, which reflects the precise positioning characteristics of the actuator. It can

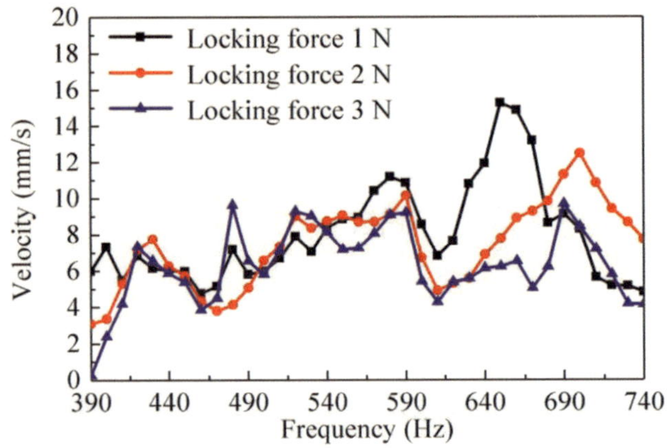

Figure 11.
Frequency characteristics under different locking forces.

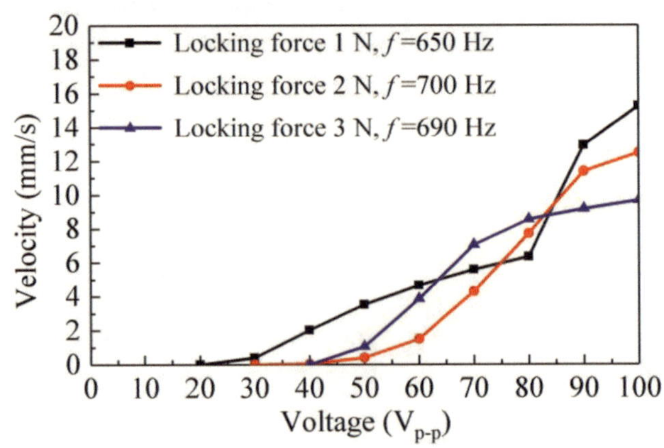

Figure 12.
Voltage characteristics under different locking forces.

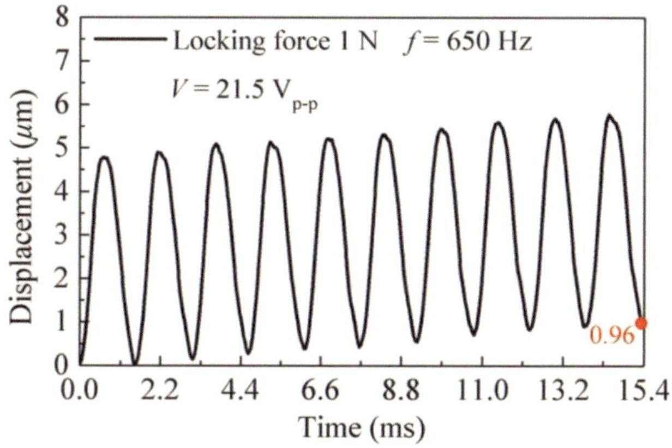

Figure 13.
The motion resolution under the locking force of 1 N.

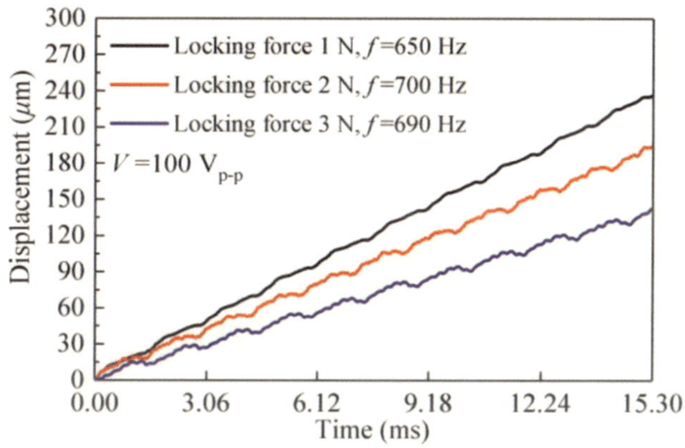

Figure 14.
Displacement characteristics under different locking forces.

be seen from **Figure 13**, the single step motion resolution of the actuator reaches 96 nm under the locking force of 1 N.

Maintaining the voltage is 100 $V_{p\text{-}p}$, the displacement characteristics under different locking forces are plotted, as shown in **Figure 14**. It is obvious that the velocity is the fastest at the locking force of 1 N, the friction resistance is small at this time, so the backward motion is minimum. The load characteristics of the actuator is emerged in **Figure 15**, as the locking force gradually increases, the maximum load of the actuator increases significantly. Within a certain adjustment range, the greater locking force can increase the friction driving force, which improves the load capacity of the actuator. It can be seen that the velocity decreases almost linearly with the load increases, and the maximum load mass of the actuator exceeds 330 g under 3 N locking force.

The efficiency η is usually introduced to evaluate the output capacity of the actuator, which can be calculated by

$$\eta = \frac{P_{\text{out}}}{P_{\text{in}}} = \frac{F \times v}{P_{\text{in}}} = \frac{mg \times v}{P_{\text{in}}} \times 100\% \tag{23}$$

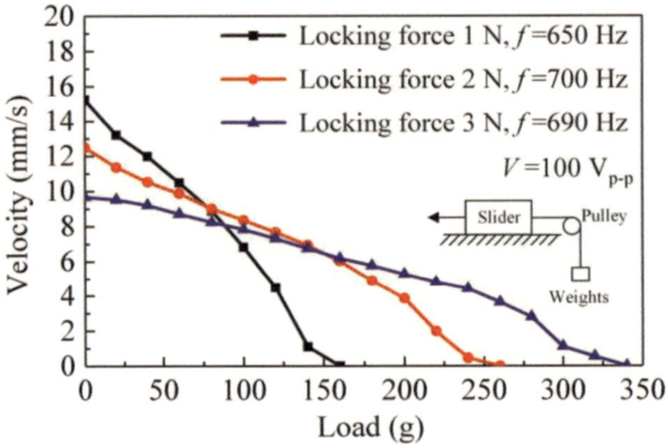

Figure 15.
Load characteristics under different locking forces.

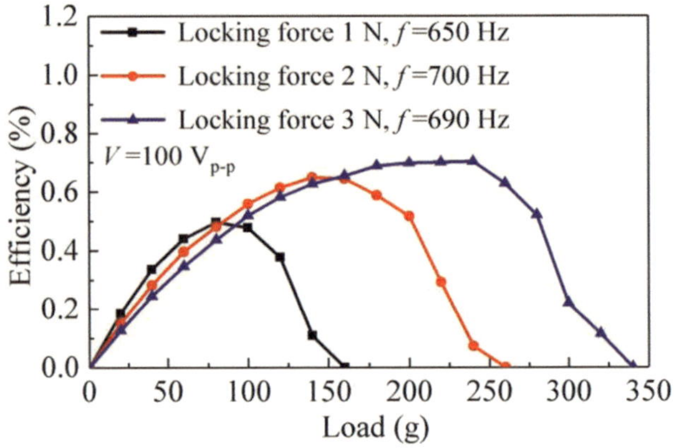

Figure 16.
Efficiencies under different locking forces.

where P_{out} represents the output power, P_{in} represents the input power, F is the gravity load, and v is the output velocity, respectively. The input power can be detected by the power analyzer, and according to the load characteristics of actuator (in **Figure 15**), the output power P_{out} can be calculated.

Figure 16 shows that the efficiency of the actuator increases first and then decreases with the increases of load. It can be seen that when the locking force is 3 N, the efficiency of the actuator reaches the highest value 0.70% under the load of 240 g.

5. Conclusions

A topology optimization work of piezoelectric actuators is illustrated to improve their output performances. As shown in this chapter, topology optimization is a powerful tool for the design of piezoelectric actuators. The aim of this work is to serve as an introduction for those who are interested in the piezoelectric actuator optimization research field, and provide a reference work for scholars. It has to be pointed out that there remain a series of unsolved problems in the field of designing piezoelectric actuators using topology optimization. In addition to the potential future work mentioned in each section, the following aspects should be considered to further promote the development of this research field: (1) The static topology optimization methods provide an efficient way to find the optimal design of piezoelectric actuators. However, some manufacturing-oriented static topology optimization methods for the design of piezoelectric actuators need to receive more attention. (2) Topology optimization of piezoelectric actuators considering dynamic problems is another important topic. When the loads changed rapidly, such as "stick–slip" type loads, the dynamic response problems of piezoelectric actuators should be concerned. (3) More topology optimization codes should be released, especially for the design of piezoelectric actuators. (4) Multi-materials topology optimization problem can be applied to piezoelectric actuators, the related theory should be considered by many outstanding scholars in the further.

Acknowledgements

This work was financially supported by the Science and Technology Development Plan of Jilin Province (Nos. 20200201057JC and 20190201108JC), and the Technology Research Planning Project of Education Department of Jilin Province (No. JJKH20220690KJ).

The authors would like to thank K. Svanberg for providing the Matlab code of MMA optimizer.

Conflict of interest

The author(s) declared no potential conflicts of interest with respect to the research, authorship, and/or publication of this article.

Author details

Shitong Yang, Yuelong Li, Guangda Qiao, Xiaosong Zhang and Xiaohui Lu*
School of Mechatronic Engineering, Changchun University of Technology, Changchun, China

*Address all correspondence to: luxh13@ccut.edu.cn

References

[1] Bendsøe MP, Kikuchi N. Generating optimal topologies in structural design using a homogenization method. Computer Methods in Applied Mechanics and Engineering. 1988;**71**: 197-224. DOI: 10.1016/0045-7825(88) 90086-2

[2] Bendsøe MP. Optimal shape design as a material distribution problem. Structural optimization. 1989;**1**:193-202. DOI: 10.1007/BF01650949

[3] Bhowmick S, Hintsala E, Stauffer D, Syed Asif SA. In-situ SEM and TEM nanomechanical study of wear and failure mechanisms. Microscopy and Microanalysis. 2018;**24**:1934-1935. DOI: 10.1017/S1431927618010152

[4] Breguet JM, Driesen W, Kaegi F, Cimprich T. Applications of Piezo-Actuated Micro-Robots in Micro-Biology and Material Science. Harbin, China: 2007 International Conference on Mechatronics and Automation; 5–8 August 2007, IEEE; 2007. pp. 57-62

[5] Wang L, Chen W, Liu J, Deng J, Liu Y. A review of recent studies on non-resonant piezoelectric actuators. Mechanical Systems and Signal Processing. 2019;**133**:106254. DOI: 10.1016/j.ymssp.2019.106254

[6] Wang S, Rong W, Wang L, Pei Z, Sun L. Design, analysis and experimental performance of a novel stick-slip type piezoelectric rotary actuator based on variable force couple driving. Smart Materials and Structures. 2017;**26**: 055005. DOI: 10.1088/1361-665X/aa64c3

[7] Liu Y, Wang L, Gu Z, Quan Q, Deng J. Development of a two-dimensional linear piezoelectric stepping platform using longitudinal-bending hybrid actuators. IEEE Transactions on Industrial Electronics. 2019;**66**: 3030-3040

[8] Tian Y, Zhang D, Shirinzadeh B. Dynamic modelling of a flexure-based mechanism for ultra-precision grinding operation. Precision Engineering. 2001; **35**:554-565

[9] Cherepanov V, Coenen P, Voigtländer B. A nano-positioner for scanning probe microscopy: The KoalaDrive. Review of Scientific Instruments. 2012;**83**:023703. DOI: 10.1063/1.3681444

[10] Kratochvil BE, Dong L, Nelson BJ. Real time rigid-body visual tracking in a scanning electron microscope. The International Journal of Robotics Research. 2009;**28**:498-511. DOI: 10.1177/0278364908099849

[11] Cheng T, He M, Li H, Lu X, Zhao H, Gao H. A novel trapezoid-type stick–slip piezoelectric linear actuator using right circular flexure hinge mechanism. IEEE Transactions on Industrial Electronics. 2017;**64**:5545-5552. DOI: 10.1109/ TIE.2017.2677318

[12] Lu X, Gao Q, Li Y, Yu Y, Zhang X, Qiao G, et al. A linear piezoelectric stick-slip actuator via triangular displacement amplification mechanism. IEEE Access. 2020;**8**:6515-6522. DOI: 10.1109/ ACCESS.2019.2963680

[13] Oubellil R, Voda A, Boudaoud M, Régnier S. Mixed stepping/scanning mode control of stick-slip SEM-integrated nano-robotic systems. Sensors and Actuators A: Physical. 2019;**285**: 258-268. DOI: 10.1016/j.sna.2018.08.042

[14] Gao X, Zhang S, Deng J, Liu Y. Development of a small two-dimensional

robotic spherical joint using a bonded-type piezoelectric actuator. IEEE Transactions on Industrial Electronics. 2021;**68**:724-733. DOI: 10.1109/TIE.2019.2959475

[15] Le T, Danh LT, Jeon JU. A proposal of a piezo rotary positioning device: Design, modeling and experiments. Smart Materials and Structures. 2019;**28**: 115032. DOI: 10.1088/1361-665X/ab4736

[16] Canfield S, Frecker M. Topology optimization of compliant mechanical amplifiers for piezoelectric actuators. Structural and Multidisciplinary Optimization. 2000;**20**:269-279. DOI: 10.1007/s001580050157

[17] Claeyssen F, Letty RL, Barillot F, Sosnicki O. Amplified piezoelectric actuators: Static and dynamic applications. Ferroelectrics. 2007;**351**: 3-14. DOI: 10.1080/00150190701351865

[18] Guo Z, Tian Y, Zhang D, Wang T, Wu M. A novel stick-slip based linear actuator using bi-directional motion of micropositioner. Mechanical Systems and Signal Processing. 2019;**128**:37-49. DOI: 10.1016/j.ymssp.2019.03.025

[19] Huang H, Zhao H, Fan Z, Zhang H, Ma Z, Yang Z. Analysis and experiments of a novel and compact 3-DOF precision positioning platform. Journal of Mechanical Science and Technology. 2013;**27**:3347-3356. DOI: 10.1007/s12206-013-0856-6

[20] Hunstig M, Hemsel T, Sextro W. Stick–slip and slip–slip operation of piezoelectric inertia drives. Part I: Ideal excitation. Sensors and Actuators A: Physical. 2013;**200**:90-100. DOI: 10.1016/j.sna.2012.11.012

[21] Li J, Zhou X, Zhao H, Shao M, Hou P, Xu X. Design and experimental performances of a piezoelectric linear actuator by means of lateral motion. Smart Materials and Structures. 2015;**24**: 065007. DOI: 10.1088/0964-1726/24/6/065007

[22] Li J, Zhou X, Zhao H, Shao M, Li N, Zhang S, et al. Development of a novel parasitic-type piezoelectric actuator. IEEE/ASME Transactions on Mechatronics. 2017;**22**:541-550. DOI: 10.1109/TMECH.2016.2604242

[23] Sigmund O. On the design of compliant mechanisms using topology optimization. Mechanics of Structures and Machines. 1997;**25**:493-524. DOI: 10.1080/08905459708945415

[24] Sigmund O. A 99 line topology optimization code written in Matlab. Structural and Multidisciplinary Optimization. 2001;**21**:120-127. DOI: 10.1007/s001580050176

[25] Bendsøe MP, Sigmund O. Topology Optimization: Theory: Methods and Applications. Berlin: Springer; 2004

[26] Da D. Topology Optimization Design of Heterogeneous Materials and Structures. Hoboken: Wiley; 2019

[27] Li Y, Li H, Cheng T, Lu X, Zhao H, Chen P. Note: Lever-type bidirectional stick-slip piezoelectric actuator with flexible hinge. Review of Scientific Instruments. 2018;**89**:086101. DOI: 10.1063/1.5038640

[28] Li J, Huang H, Morita T. Stepping piezoelectric actuators with large working stroke for nano-positioning systems: A review. Sensors and Actuators A: Physical. 2019;**292**:39-51. DOI: 10.1016/j.sna.2019.04.006

[29] Hunstig M. Piezoelectric inertia motors: A critical review of history, concepts, design, applications, and

perspectives. Actuators. 2017;**6**:7. DOI: 10.3390/act6010007

[30] Iqbal S, Malik A. A review on MEMS based micro displacement amplification mechanisms. Sensors and Actuators A: Physical. 2019;**300**:111666. DOI: 10.1016/j.sna.2019.111666

[31] Kang D, Lee MG, Gweon D. Development of compact high precision linear piezoelectric stepping positioner with nanometer accuracy and large travel range. Review of Scientific Instruments. 2007;**78**:075112. DOI: 10.1063/1.2756627

[32] Kim JH, Kim SH, Kwak YK. Development of a piezoelectric actuator using a three-dimensional bridge-type hinge mechanism. Review of Scientific Instruments. 2003;**74**:2918. DOI: 10.1063/1.1569411

[33] Lau GK, Du HJ, Guo NQ, Lim MK. Systematic design of displacement-amplifying mechanisms for piezoelectric stacked actuators using topology optimization. Journal of Intelligent Material Systems and Structures. 2000; **11**:685-695. DOI: 10.1106/17V9-DR0Q-6BMU-RKUY

[34] Lee HJ, Kim HC, Kim HY, Gweon DG. Optimal design and experiment of a three-axis out-of-plane nano positioning stage using a new compact bridge-type displacement amplifier. Review of Scientific Instruments. 2013;**84**:115103. DOI: 10.1063/1.4827087

[35] Yang S, Li Y, Xia X, Ning P, Ruan W, Zheng R, et al. A topology optimization method and experimental verification of piezoelectric stick-slip actuator with flexure hinge mechanism. Archive of Applied Mechanics. 2022;**92**: 271-285. DOI: 10.1007/s00419-021-02055-4

[36] Ling M, Cao J, Zeng M. Enhanced mathematical modeling of the displacement amplification ratio for piezoelectric compliant mechanisms. Smart Materials and Structures. 2016;**25**: 075022. DOI: 10.1088/0964-1726/25/7/075022

[37] Liu K, Tovar A. An efficient 3D topology optimization code written in Matlab. Structural and Multidisciplinary Optimization. 2014;**50**:1175-1196. DOI: 10.1007/s00158-014-1107-x

[38] Schlinquer T, Mohand-Ousaid A, . Rakotondrabe M. Displacement Amplifier Mechanism for Piezoelectric Actuators Design Using SIMP Topology Optimization Approach. In: 2018 IEEE International Conference on Robotics and Automation (ICRA); 21–25 May 2018; Brisbane, QLD, Australia: IEEE; 2018. p. 4305–4311

[39] Yang S, Xia X, Liu X, Qiao G, Zhang X, Lu X. Improving velocity of stick-slip piezoelectric actuators with optimized flexure hinges based on SIMP method. IEEE Access. 2020;**8**: 213122-213129. DOI: 10.1109/ACCESS.2020.3039857

[40] Zhang X, Zhu B. Topology Optimization of Compliant Mechanisms. Singapore: Springer; 2018

[41] Svanberg K. The method of moving asymptotes—A new method for structural optimization. International Journal for Numerical Methods in Engineering. 1987;**24**: 359-373. DOI: 10.1002/nme.1620240207

[42] Svanberg K. MMA and GCMMA—Two Methods for Nonlinear Optimization [Internet]. 2007. Available from: https://people.kth.se/~krille/mmagcmma.pdf. [Accessed: May 2, 2022]

Chapter 7

Active Vibration Suppression Based on Piezoelectric Actuator

Min Wang, Songquan Liao, Xuan Fang and Shibo Fu

Abstract

The piezoelectric constitutive equation states that the inverse piezoelectric effect can convert electrical energy into mechanical energy, resulting in small displacement and force changes with high resolution. The piezoelectric actuator based on inverse piezoelectric effect has an excellent performance in active vibration suppression because of its high frequency response, high positioning accuracy, and large output force. A new active-passive composite vibration suppression system can be formed by cascading it with passive vibration isolation elements in series and parallel. On this basis, by adding different control algorithms and control loops, such as the Sky-Hook damping feedback control algorithm and adaptive feedforward control algorithm, different vibration control effects can be realized.

Keywords: vibration isolation, active control, active-passive, inverse piezoelectric effect, piezoelectric actuator

1. Introduction

Mechanical vibration exists in various machines in working conditions, such as precision machine tools, aircrafts, ships, etc. Strong vibration will affect the accuracy and stability of mechanical parts. In severe cases, it will also lead to fatigue failure and shorten the life of the structure or cause resonance to damage the structure. Therefore, suppressing the unfavorable vibration response has become an urgent problem to be solved in the industry.

The passive vibration suppression system commonly used in the engineering field achieves the purpose of vibration suppression by installing elastic damping elements to consume and absorb vibration energy. The system has high reliability but cannot adjust the vibration suppression characteristics and cannot adapt to changes in the external environment. Therefore, the active vibration suppression system with active adjustment capability and wide adaptive frequency range has become a research hotspot.

The active vibration suppression system is composed of sensors, actuators, and control systems, so the current research mainly focuses on two directions. One is the study of active control strategies; the other is the study of new materials and corresponding actuators. The active control strategies currently used in the field of

IntechOpen

vibration suppression include PID control [1], adaptive control [2, 3], intelligent control [4, 5], and so on. The actuator is one of the key elements that affect the vibration suppression performance of the entire active vibration suppression system. In recent years, the development of smart material structures has provided conditions for the development of actuators, which has greatly promoted the research and application of active vibration control technology. At present, the smart materials used in the design of active vibration suppression actuators mainly include electro/magnetorheological fluids [6, 7], shape memory alloys [8, 9], magnetostrictive materials [10, 11], piezoelectric materials [12], and so on. Among them, the piezoelectric materials can be used as both actuators (inverse piezoelectric effect) and sensors (positive piezoelectric effect) due to their positive and inverse piezoelectric effects. As actuators, they have many advantages such as fast frequency response, wide control frequency range, high displacement resolution, small size, easy integration, no mechanical friction, and so on [13]. They have been widely used in active vibration suppression systems.

Most of the active suppression methods based on piezoelectric actuators reduce the vibration by directly suppressing the excitation force. That is, when the piezoelectric actuators are arranged, the direction of the output force/displacement of the actuators is consistent with the direction of the system vibration. In order to improve the vibration suppression effect, it is usually required that the piezoelectric actuator can output a sufficiently large displacement and control force at the same time, but its realization is limited by the electromechanical coupling characteristics of the piezoelectric actuator. This chapter discusses an active-passive composite vibration suppression system based on piezoelectric actuators. The active control element adopts a piezoelectric stack actuator with a mechanical displacement amplifying mechanism. The piezoelectric stack actuator has the advantages of high energy conversion efficiency, low operating voltage, and large output force. The mechanical displacement amplifying mechanism has a compact structure and can effectively amplify the displacement of the actuator. The second section of the chapter will analyze the driving characteristics of the piezoelectric actuator and the magnification of the displacement amplifying mechanism in the active control element, and discuss the compensation of hysteresis that affects the control accuracy of the piezoelectric actuator. Section 3 will analyze and construct the structure of the active-passive composite vibration suppression system based on the piezoelectric actuator, and establish its dynamic model. Section 4 will analyze the vibration control theory of the active-passive composite vibration suppression system using different control algorithms and control loops on the basis of the system dynamics model, and simulate the effectiveness of the algorithms. Section 5 will build an experimental platform to verify the active vibration suppression effect of the active-passive composite vibration suppression system based on the piezoelectric actuator. Section 6 is the conclusion, which will summarize the content of this chapter.

2. Characteristics of piezoelectric actuator in vibration suppression

2.1 Inverse piezoelectric effect of piezoelectric actuator

The piezoelectric constitutive equation states that the inverse piezoelectric effect of piezoelectric actuators can convert electrical energy into mechanical energy, resulting in small displacement and force changes with high resolution.

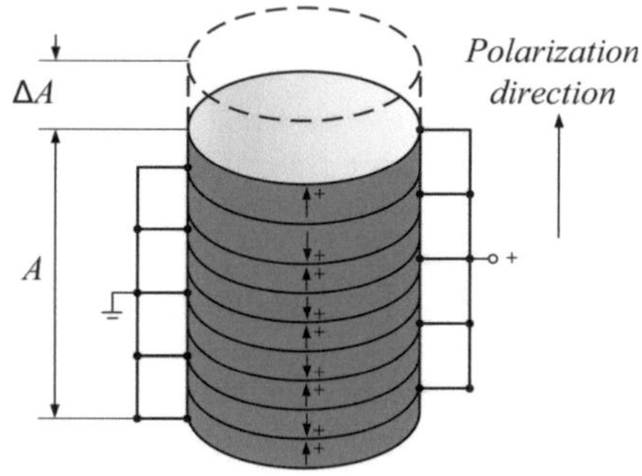

Figure 1.
Structure diagram of piezoelectric stack actuator.

The piezoelectric stack actuator used in this chapters is formed by stacking many thin piezoelectric ceramic sheets, and its structure is shown in **Figure 1**. Each piezoelectric ceramic sheet is equipped with electrodes and is separated by an insulating layer. The elongation of the piezoelectric ceramic sheet along the polarization direction is mainly related to the applied electric field strength but has nothing to do with its thickness. With the layered stack structure, the elongation of the piezoelectric ceramic sheets can be accumulated, so that the piezoelectric stack actuator can still generate a relatively large displacement at a lower operating voltage.

According to the inverse piezoelectric effect of piezoelectric materials, only considering the longitudinal force and elongation of the piezoelectric ceramic sheet under the action of the driving voltage, and assuming that the strain is uniformly distributed in the longitudinal polarization direction, then the longitudinal stress of a single piezoelectric ceramic sheet can be expressed as:

$$\sigma = E_p(\varepsilon - d_{33}E) \tag{1}$$

where σ is the longitudinal stress; E_p is the initial elastic modulus; ε is the longitudinal strain; d_{33} is the longitudinal piezoelectric strain constant; E is the longitudinal electric field strength.

Assuming that the force-bearing area of the piezoelectric ceramic sheet is A_p, the thickness is h, the voltage applied at both ends is V and the resulting thickness deformation is δ. Then the longitudinal strainε and longitudinal electric field strength E in Eq. (1) can be expressed as:

$$\varepsilon = \delta/h \tag{2}$$

$$E = V/h \tag{3}$$

The force on the piezoelectric ceramic sheet can be expressed as the stress σ multiplied by the force-bearing area A_p, and the relationship between the output force and the force on the piezoelectric ceramic sheet is force and reaction force. Combining Eqs. (1–3), the output force of the piezoelectric ceramic sheet is:

$$f = -\sigma A_p = A_p E_p/h(d_{33}V - \delta) \tag{4}$$

Let $k = A_p E_p/h$, which is the inherent constant of the piezoelectric ceramic sheet, the output force expression of the piezoelectric ceramic sheet can be simplified as:

$$f = k(d_{33}V - \delta) \tag{5}$$

According to the output force expression Eq. (5), the output displacement of the piezoelectric ceramic sheet can be obtained as:

$$\delta = d_{33}V - f/k \tag{6}$$

Since the piezoelectric stack actuator cascades several piezoelectric ceramic sheets in voltage parallel and physical series, its output force is the same as that of a single piezoelectric ceramic sheet, and its output displacement is the sum of the output displacement of all piezoelectric ceramic sheets:

$$\begin{cases} f_z = f \\ \Delta A = n\delta \end{cases} \tag{7}$$

where f_z is the output force of the piezoelectric actuator, ΔA is the output displacement of the piezoelectric actuator, and$n$ is the number of piezoelectric ceramic sheets in the piezoelectric actuator.

Although the output displacement of the piezoelectric stack actuator is the sum of the output displacements of all piezoelectric ceramic sheets, its stroke is still in the order of microns. In practical applications, it is usually necessary to cooperate with a displacement amplifying mechanism [14–16]. Therefore, it is necessary to analyze the displacement amplification characteristics of the displacement amplifying mechanism.

2.2 Displacement amplifying mechanism of piezoelectric actuator

The mechanical displacement amplifying mechanism adopted in this chapter is based on the principle of triangular amplification to mechanically amplify the displacement of the piezoelectric actuator, to make up for the shortcoming of its insufficient stroke and expand its effective stroke. Dimensions such as the coordinate direction and angle of the displacement amplifying mechanism are shown in **Figure 2a**. The piezoelectric stack actuator with length A is placed in the x-direction of the displacement amplifying mechanism, and the inclination angle formed by the horizontal direction and the hypotenuse of the amplifying mechanism is α.

The amplification principle of the mechanical displacement amplifying mechanism is shown in **Figure 2b**. When the displacement changes in the x-direction, the inverse change of the displacement occurs in the y-direction. The displacement magnification

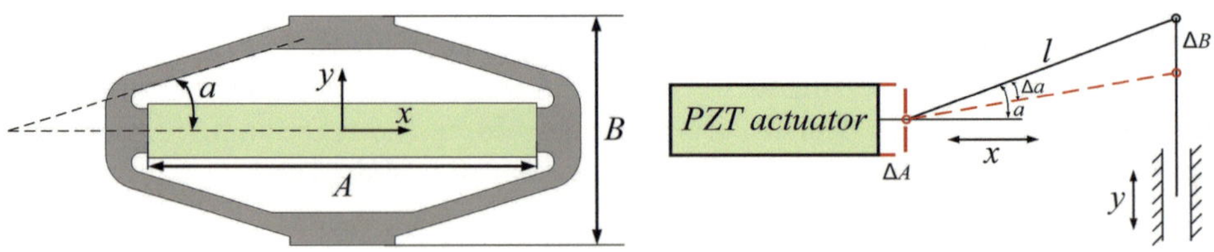

(a) (b)

Figure 2.
Displacement amplifying mechanism: (a) dimension labeling diagram; (b) amplifying schematic diagram.

is defined as the ratio of the displacement change in the y-direction to the displacement change in the x-direction, which can be expressed as:

$$\gamma = \frac{\Delta B}{\Delta A} \tag{8}$$

According to the geometric transformation relationship shown in **Figure 2b**, we can get:

$$\begin{aligned} \Delta A &= l\cos\alpha - l\cos(\alpha - \Delta\alpha) \\ \Delta B &= l\sin\alpha - l\sin(\alpha - \Delta\alpha) \end{aligned} \tag{9}$$

Then the displacement magnification can be organized as:

$$\gamma = \frac{\Delta B}{\Delta A} = \frac{l\sin\alpha - l\sin(\alpha - \Delta\alpha)}{l\cos\alpha - l\cos(\alpha - \Delta\alpha)} = \frac{\tan\alpha(1 - \cos(\Delta\alpha)) + \sin(\Delta\alpha)}{1 - \cos(\Delta\alpha) - \tan\alpha\sin(\Delta\alpha)} \tag{10}$$

Since the stroke of the piezoelectric actuator is in the order of microns, the variation in length ΔA and the variation in inclination angle $\Delta\alpha$ are quite small. According to the equivalent infinitesimal principle: $\sin(\Delta\alpha)\approx\Delta\alpha$, $1 - \cos(\Delta\alpha)\approx\frac{1}{2}(\Delta\alpha)^2$, $\tan(\Delta\alpha)\approx\Delta\alpha$. The Eq. (10) can be further organized as:

$$\gamma = \frac{\Delta\alpha\tan\alpha + 2}{\Delta\alpha - 2\tan\alpha} \approx -\frac{1}{\tan\alpha} \tag{11}$$

It can be seen from Eq. (11) that the magnification of the triangular mechanical displacement amplifying mechanism is not related to the length but is only related to the size of the inclination angle α. The negative sign indicates that the displacement changes in the y- and x-directions are opposite. The relationship between magnification and inclination angle is described in **Figure 3**.

It can be seen from **Figure 3** that the smaller the inclination angle, the larger the magnification (regardless of positive and negative). As the inclination angle increases, the change in magnification becomes insignificant. When the inclination angle reaches 45 degrees, the magnification is close to 1.

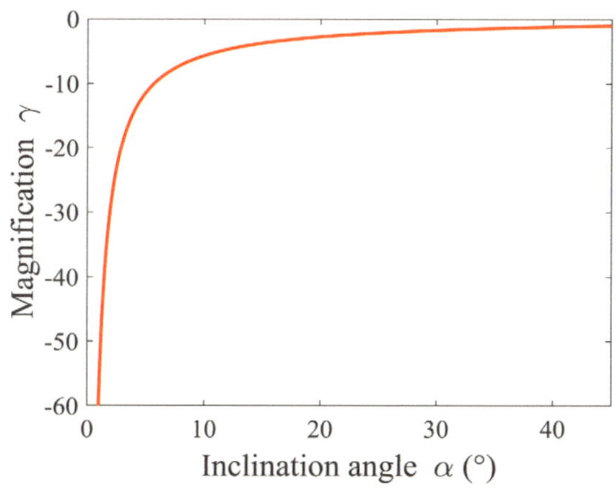

Figure 3.
Relationship between magnification and inclination angle.

2.3 Compensation of hysteresis in piezoelectric actuator

According to Eqs. (6–7), under ideal conditions, the output displacement of the piezoelectric actuator is proportional to the driving voltage, as shown by the red solid line in **Figure 4**. But, in fact, due to the difference in structure and the complexity of its working mechanism, piezoelectric actuators have the inherent characteristics of hysteresis [17], which leads to the fact that the voltage rise and drop curve of the output displacement do not overlap, as shown by the green and blue lines in **Figure 4**. This affects the control accuracy of the piezoelectric actuator to a certain extent. In general engineering applications, the displacement error is relatively small, and the piezoelectric actuator can be used approximately linearly. However, in some applications that require high control accuracy, it is necessary to perform hysteresis compensation [18–21]. The hysteresis in the piezoelectric actuator can be compensated by the hysteresis inverse compensation method. The specific compensation control block diagram is shown in **Figure 5**. This method can avoid complex modeling and parameter identification of the piezoelectric actuator, and the error output can be directly used for compensation control to achieve the effect of eliminating hysteresis and improve the control accuracy of the piezoelectric actuator.

3. Active-passive composite vibration suppression system based on piezoelectric actuator

The passive vibration suppression system has a simple structure and good high-frequency vibration suppression performance, but it is powerless to low-frequency vibration, and the resonance peak suppression conflicts with high-frequency vibration suppression. The vibration suppression performance of the active vibration suppression system is good, but due to the limitation of the working bandwidth and power of the actuator, it is difficult to realize the active vibration suppression only. Therefore, active actuators are usually used in combination with passive elements to form an active-passive composite vibration suppression system to achieve the best vibration suppression effect [22–24].

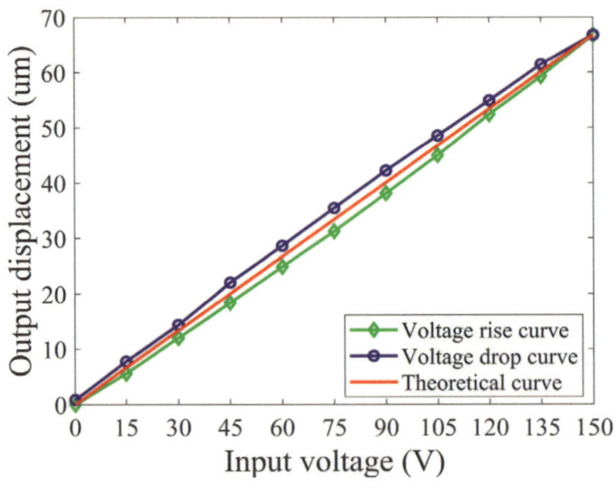

Figure 4.
The relationship between voltage and displacement of piezoelectric actuator.

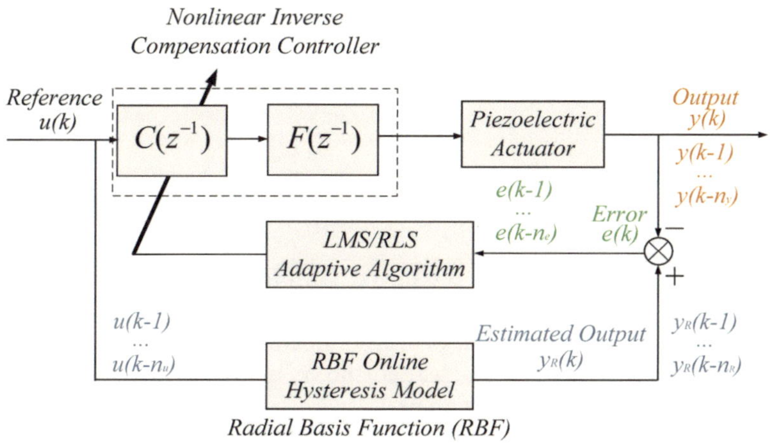

Figure 5.
Hysteresis adaptive inverse compensation control block diagram of piezoelectric actuator.

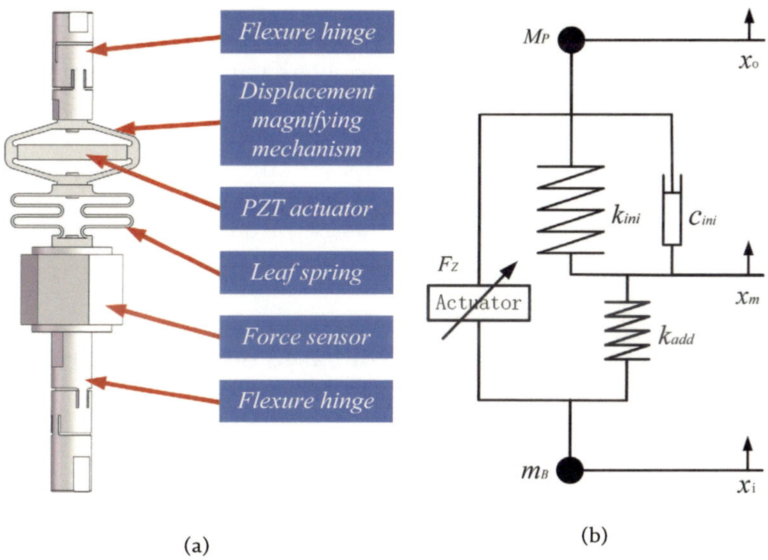

Figure 6.
Single-degree-of-freedom vibration suppression system: (a) three-dimensional model diagram; (b) simplified model diagram.

The active control element based on the piezoelectric actuator has high stiffness. In order to reduce the overall stiffness of the vibration suppression system, thereby reducing the natural frequency of the system and increasing the vibration suppression bandwidth, the active control element and the leaf spring with relatively low stiffness are connected in series to form a vibration suppression system with an active-passive composite structure in this paper, as shown in **Figure 6a**. The vibration suppression system is a single-degree-of-freedom system. The modular design reduces the mass and size of the system to the greatest extent, which makes it more suitable for applications in those fields with strict space and mass constraints, such as aerospace, ships, etc. In addition, according to the needs of the application, it can be conveniently used as the branch chain of the multi-degree-of-freedom vibration suppression platform through the flexible hinges [13, 25–27]. Its simplified model is shown in **Figure 6b**.

According to Newton's second law, the dynamic equation of the system can be obtained as:

$$\begin{cases} M_p\ddot{x}_o + c_{ini}(\dot{x}_o - \dot{x}_m) + k_{ini}(x_o - x_m) = F_z \\ c_{ini}(\dot{x}_o - \dot{x}_m) + k_{ini}(x_o - x_m) = k_{add}(x_m - x_i) \end{cases} \tag{12}$$

where M_p is the load mass, c_{ini} is the initial damping of the system, k_{ini} is the initial stiffness of the system, k_{add} is the additional stiffness of the leaf spring, x_o is the load displacement, x_m is the displacement of the connection point between the displacement amplifying mechanism and the leaf spring, x_i is the base displacement, F_z is the output force of the active control element.

By combining the two equations in Eq. (12) and eliminating the relevant variables at the intermediate connection point, the dynamic model of the single-degree-of-freedom vibration suppression system is obtained as follows:

$$M_p\ddot{x}_o + c(\dot{x}_o - \dot{x}_i) + k_d(x_o - x_i) = F_z \tag{13}$$

where $c = k_{add}c_{ini}/(k_{add} + k_{ini})$ is the equivalent damping of the system and $k_d = k_{add}k_{ini}/(k_{add} + k_{ini})$ is the equivalent stiffness of the system.

It can be seen from Eq. (13) that the active control element in series with the leaf spring can effectively reduce the overall stiffness of the vibration suppression system, and the establishment of the structure and dynamic model of the active-passive composite vibration suppression system provides the object and theoretical basis for the subsequent active vibration suppression control.

4. Active vibration suppression control based on piezoelectric actuator

4.1 Piezoelectric active vibration suppression system

Passive vibration suppression refers to the introduction of one or more mass-spring damping systems in the propagation path of the vibration source. Although this technical solution is simple and reliable, it can only effectively attenuate high-frequency vibrations in a wide frequency band. With the rapid development of smart sensors and smart actuators and high-speed microprocessors, active vibration suppression is becoming more and more attractive in vibration suppression. In piezoelectric active control, it is classified according to the control structure, which can be divided into feedforward control and feedback control. In precision vibration isolation, different vibration active control structures need to be adopted for different vibration isolation objects in actual work.

In the piezoelectric active vibration suppression system, the overall control block diagram of feedforward and feedback is shown in **Figure 7**. In **Figure 7**, The vibration suppression closed-loop consists of a table feedback loop with a signal filter function and a ground-based feedforward loop with a signal filter function. The feedback loop is implemented as: The feedback acquisition sensor collects the table vibration, and then filters the excess noise signal, and then transmits it to the feedback controller for algorithm calculation, and adjusts the gain of the entire feedback loop to change the feedback characteristics. The implementation of the feedforward loop is as follows:

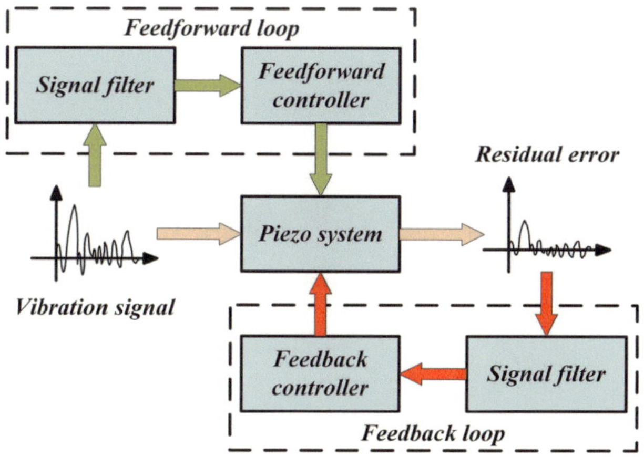

Figure 7.
Piezoelectric active vibration suppression control block diagram.

The feedforward acquisition sensor collects the ground vibration signal, then filters the excess noise signal, and then transmits it to the feed-forward controller for algorithm calculation, and adjusts the feedforward control parameters to change the feedforward characteristics. The piezoelectric feedback active control algorithm can be realized by sky-hook damping, integral force feedback (IFF), and other algorithms, and the feedforward control algorithm can be realized by adaptive control algorithm or phase compensation algorithm. Finally, the hybrid operation of the two active controllers is output to the piezoelectric actuator, and the active vibration suppression system is controlled to complete the active control. Compared with using one of the control structures or algorithms alone, the active hybrid control (AHC) can achieve better vibration suppression performance. The following will introduce a piezoelectric active hybrid control design method, including an IFF control and a recursive least square (RLS) adaptive feedforward control, the feedback control realizes the sky-hook damping effect, and the adaptive feedforward control realizes the ground-based advance response.

4.2 Piezoelectric IFF control in vibration suppression

Active feedback control can effectively solve the problem that the signal at the natural frequency is amplified, that is, the problem of formant attenuation. At present, the sky-hook technology is widely used in piezoelectric active feedback control. Generally, the sky-hook effect is established by absolute speed feedback control to reduce the formant peak value while maintaining high-frequency attenuation. This section introduces sky-hook technology based on piezoelectric actuators, which uses a combination of dynamic force sensors and piezoelectric actuators to design an integral force active control algorithm to achieve the control effect of sky-hook.

According to the structure above, the schematic diagram of the piezoelectric IFF control is shown in **Figure 8**. In **Figure 8**, the amount of elongation of the piezoelectric actuator is represented by δ, F represents the dynamic force signal. The active vibration control method of piezoelectric IFF control is: The dynamic force sensor collects the dynamic force signal, and after noise removal and filtering, the integral

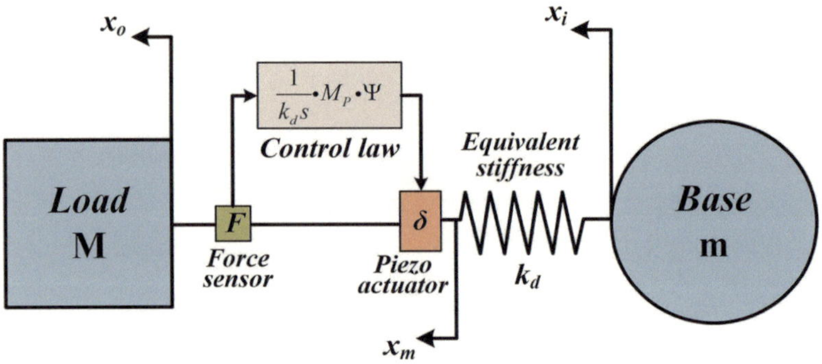

Figure 8.
Schematic diagram of piezoelectric IFF principle.

calculation compensation is completed in the active control unit, and the control signal is output to the piezoelectric actuator to complete the piezoelectric active feedback control.

The IFF control law based on sky-hook damping technology is:

$$\sigma_{IFF} = \frac{1}{k_d s} \cdot \Psi \cdot F = \frac{1}{k_d s} \cdot M_P \cdot \Psi \cdot s^2 = \frac{1}{k_d s} \cdot k_i \cdot s^2 \tag{14}$$

where $k_i = M_P \cdot \Psi$ is the integral gain coefficient of the IFF control.

According to **Figure 8**, the motion control equation of the piezoelectric IFF control system can be expressed as:

$$Ms^2 x_o = -ms^2 x_i = k_d(x_i - x_m) = F \tag{15}$$

$$\delta = x_o - x_m \tag{16}$$

The open-loop transfer function between the elongation δ of the piezoelectric actuator and the output F of the force sensor can be expressed as:

$$\frac{F}{\delta} = k_d \frac{Mms^2}{Mms^2 + k_d(M+m)} \tag{17}$$

According to the integral gain coefficient of the IFF algorithm in Eq. (14), after sorting and calculation, the displacement x_m of the middle section can be obtained as:

$$x_m = \frac{sx_o + k_i x_i}{s + k_i} = \frac{sx_o + M_P \cdot \Psi \cdot x_i}{s + M_P \cdot \Psi} \tag{18}$$

According to the above derivation, the transmissibility curve of the piezoelectric vibration isolation system under the IFF control algorithm is:

$$T_{C-IFF}(s) = \frac{cs + k_d}{M_p s^2 + (c + k_i)s + k_d} = \frac{1}{s^2(1/\omega_n^2) + s(\Psi/\omega_n^2) + 1} \tag{19}$$

$$\omega_n = \sqrt{\frac{k_d}{M_p}} \tag{20}$$

where ω_n is the natural frequency of the passive vibration isolation system. When the gain factor $\Psi \gg c$, the equivalent damping *c* of the system can be ignored.

The natural frequency and damping ratio of the piezoelectric vibration isolation system under the IFF control algorithm are expressed as:

$$\omega_P = \sqrt{\frac{k}{M_P}} = \omega_n \tag{21}$$

$$\zeta_P = \frac{c + k_i}{2\sqrt{km}} = \frac{M_P \cdot \Psi}{2\sqrt{km}} = \frac{\Psi}{2}\sqrt{\frac{M_P}{k}} = \frac{\Psi}{2\omega_n} \tag{22}$$

The natural frequency of the vibration isolation system under IFF control is consistent with the natural frequency of the passive system and does not change. The damping ratio of the vibration isolation system under the IFF control is proportional to the integral gain coefficient. By increasing the integral gain coefficient, the formant peak value at the natural frequency can be effectively reduced to achieve the sky-hook effect. It is worth noting that an excessively large gain coefficient will lead to system stability errors, making the vibration isolation system unstable.

According to the theoretical analysis of IFF, the simulation analysis is carried out in Matlab. The passive system parameters are shown in **Table 1**, and the parameters of the subsequent simulation are also consistent with **Table 1**. The simulation results are shown in **Figure 9**. It can be found that with the increase of the integral gain coefficient, the value of the resonance peak of the vibration isolation system decreases continuously, which plays a good role in suppressing vibration. This shows that the piezoelectric IFF control can achieve the effect of sky-hook damping control and can effectively suppress the formant.

4.3 Piezoelectric RLS adaptive feedforward control in vibration suppression

The most direct way to improve the performance of feedback control is to increase its feedback gain. However, with the increase of the feedback gain, a large steady-state error will be introduced into the system. Therefore, a ground-based feedforward control strategy emerges as the times require. The ground-based feedforward control can effectively improve the local frequency-domain vibration suppression performance of the system by predicting the vibration signal in advance and implementing active control in the active algorithm. In piezoelectric feedforward control, the use of adaptive feedforward control is an extremely effective method. This section introduces an RLS adaptive feedforward control method.

The adaptive controller $F_{ff} = F[y_i(k), d(k), y_o(k)]$ is a finite impulse response filter (FIR), also known as a transversal filter. For the observation signal that changes with time *i*, the tap weight vector w(n) of the transversal filter must be time-varying.

Simulation type	Passive Parameters
Load Mass (Kg)	1
System Stiffness (N/m)	3.41e4
System Damping (N.m/s)	7.65

Table 1.
Simulation parameters of single-degree-of-freedom passive vibration isolation system.

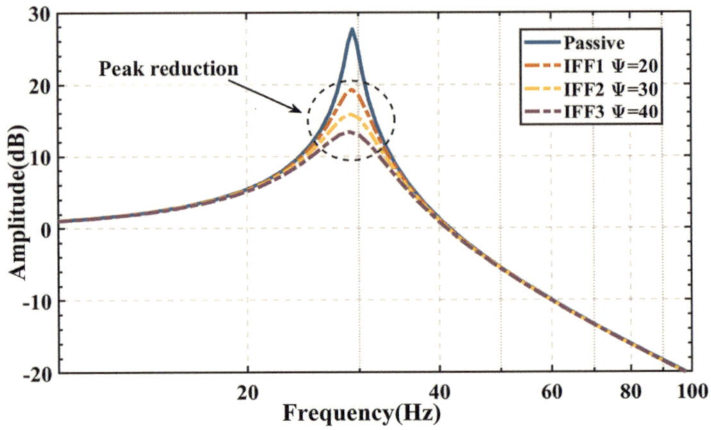

Figure 9.
Transmittance curve of piezoelectric vibration isolation system under passive control and IFF control.

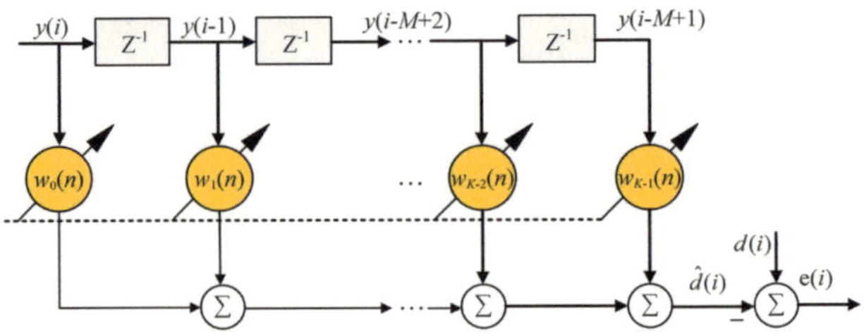

Figure 10.
Transverse filter with time-varying tap weights.

To keep the adaptive controller in an optimal state, that is, to keep the gradient function of the cost function close to zero, the variable parameters must converge to the optimal value in real-time. Therefore, the filter used in the RLS algorithm is a transversal filter with time-varying tap weights. **Figure 10** shows the structure diagram of the transversal filter with time-varying tap weights.

The RLS adaptive control algorithm is a transversal filter based on the least-squares criterion. The algorithm recursively deduces the weight vector of the current time according to the filter tap weight vector of the previous time. Assuming that N data $y(1), y(2), \ldots, y(i), \ldots, y(N)$ are known, the data is filtered with an M-order transversal filter with time-varying tap weights to get estimate the desired signals $d(1), d(2), \ldots, d(i), \ldots, d(N)$. Then the estimate of the expected response can be expressed as:

$$\hat{d}(i) = \sum_{j=0}^{M-1} w_j(n) y(i-j) = \boldsymbol{w}^T(n) y(i) \tag{23}$$

where $w_j(n)$ is the tap weight of the M-order transversal filter with time-varying tap weights, $\boldsymbol{w}(n)$ is the tap weight vector of the filter with time-varying tap weights, and $\boldsymbol{y}(i)$ is the tap input vector of the filter with time-varying tap weights at the i-th time, and are respectively:

$$\boldsymbol{w}(n) = [w_0(n), w_1(n), \ldots w_{M-1}(n)]^T \tag{24}$$

$$\mathrm{y}(\mathrm{i}) = [\mathrm{y}(\mathrm{i}), \mathrm{y}(\mathrm{i}\text{-}1), \ldots, \mathrm{y}(\mathrm{i}-\mathrm{M}+1)]^{\mathrm{T}} \tag{25}$$

Then the estimated error of the filter with time-varying tap weights can be written as:

$$\mathrm{e}(\mathrm{i}) = \mathrm{d}(\mathrm{i}) - \hat{\mathrm{d}}(\mathrm{i}) = \mathrm{d}(\mathrm{i}) - \sum_{\mathrm{j}=0}^{\mathrm{M}-1} \mathrm{w_j}(\mathrm{n})\mathrm{y}(\mathrm{i}-\mathrm{j}) = \mathrm{d}(\mathrm{i}) - \boldsymbol{w}^{\mathrm{T}}(\mathrm{n})\boldsymbol{y}(\mathrm{i}) \tag{26}$$

Then the cost function under the least-squares criterion using the pre-windowing method can be expressed as:

$$\xi(n) = \sum_{\mathrm{i}=1}^{\mathrm{n}} \lambda^{\mathrm{n}-\mathrm{i}} \mathrm{e}^2(\mathrm{i}) \tag{27}$$

where λ is the forgetting factor, the value range is $0 \le \lambda \le 1$.

The RLS adaptive feedforward controller is built-in Matlab for simulation, and the simulation results are shown in **Figure 11**. Under the action of piezoelectric RLS adaptive feedforward control, the effective suppression rate of active control to amplitude can reach 80%, which is obviously better than passive control.

4.4 Piezoelectric active hybrid controller

After the above description of IFF control and RLS adaptive feedforward control, a design scheme of piezoelectric AHC can be given. The block diagram of the AHC is shown in **Figure 12**. The principle is as follows: Given an additional external excitation signal, the signal $y(n)$ is measured with a feedforward sensor. Design a transversal filter $C(z^{-1})$ with time-varying tap weights, and continuously estimate the expected response $d(i)$ by fitting the tap weight vector $\boldsymbol{w}(\mathrm{n})$ constantly changing. By iteratively deriving the least squares estimated tap weight vector $\boldsymbol{w}(\mathrm{n})$, the square weighted sum of the estimated error $e(i)$ (that is, the platform vibration error of the load platform) under this system is obtained to be the smallest, thereby, ensuring the platform vibration error of the load platform. The feedforward control loop finally generates

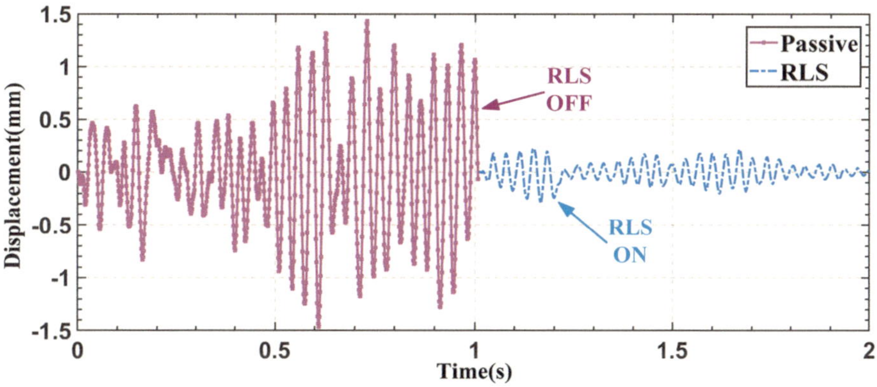

Figure 11.
Time domain comparison of RLS adaptive feedforward control and passive control.

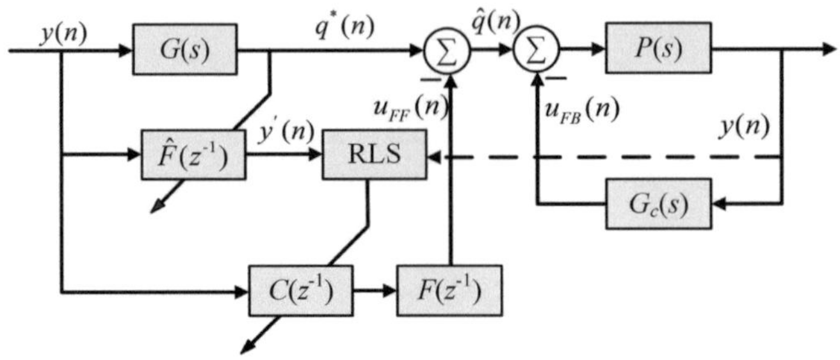

Figure 12.
Piezoelectric AHC block diagram.

the feedforward control signal $u_{FF}(\mathrm{n})$, which is converted by the piezoelectric brake into an actual force acting on the system, which is opposite to the direct interference generated by the feedforward, thereby, eliminating the system vibration caused by the feedforward source. The feedback control loop performs compensation control on the error signal according to the feedback controller.

As shown in **Figure 12**, $G(\mathrm{s})$ is the structural open-loop transfer function, $P(\mathrm{s})$ is the structural closed-loop transfer function, $G_c(\mathrm{s})$ is the feedback controller, $C(z^{-1})$ is a transversal filter, $F(z^{-1})$ is the transfer function from the feedforward control force to the vibration isolation system, $\hat{F}(z^{-1})$ is the model of $F(z^{-1})$ function with M-order identification, $y(n)$ is an additional external stimulus, $q^*(n)$ is the vibration response of the external excitation signal to the open-loop transfer function $G(\mathrm{s})$ of the structure, $u_{FF}(n)$ is the feedforward control output force, $u_{FB}(n)$ is the feedback control output force, $\hat{q}(\mathrm{n})$ is the first response error, and is the residual vibration error of the table after feedforward control, $q(\mathrm{n})$ is the second error response, and is the residual vibration error of the table after entering the hybrid control.

According to Eq. (26), combined with the AHC block diagram, the system error $e'(n)$ can be equivalent to the first residual vibration error response $\hat{q}(n)$, that is, the table residual vibration error after feedforward control:

$$e'(\mathrm{n}) = \hat{\mathrm{q}}(\mathrm{n}) = \mathrm{q}^*(\mathrm{n}) - \mathrm{u_{FF}}(\mathrm{n}) = \mathrm{q}^*(\mathrm{n}) - \boldsymbol{w}^{\mathrm{T}}(\mathrm{n})\boldsymbol{y}(\mathrm{n}) \tag{28}$$

In the real-time control of the active vibration system, due to the real-time control operation of the system, the sensor can only measure the vibration signal of the ground foundation and the vibration signal after the active hybrid control of the table, and cannot directly measure the vibration response $q^*(n)$ from the ground vibration signal to the open-loop transfer function $G(s)$. Therefore, the signal transfer can be estimated by the model reference function $\hat{F}(z^{-1})$:

$$\mathrm{q}^*(\mathrm{n}) \doteq \mathrm{y}'(\mathrm{n}) = \mathrm{y}(\mathrm{n})\hat{\mathrm{F}}(\mathrm{z}^{-1}) \tag{29}$$

Then the system expansion error $e(n)$ can be equivalent to the second error response $q(n)$, that is, the residual vibration error of the table after AHC:

$$\mathrm{e}(\mathrm{n}) = \mathrm{q}(\mathrm{n}) = \mathrm{e}'(\mathrm{n})\frac{\mathrm{P}(s)}{1 + \mathrm{P}(s)\mathrm{G_c}(s)} = \hat{\mathrm{q}}(\mathrm{n})\frac{\mathrm{P}(s)}{1 + \mathrm{P}(s)\mathrm{G_c}(s)} \tag{30}$$

According to Eq. (27), the cost function after using AHC is:

$$\xi(\mathrm{n}) = \sum_{\mathrm{i}=1}^{\mathrm{n}} \lambda^{\mathrm{n-i}} \mathrm{e}^2(\mathrm{i}) = \sum_{\mathrm{i}=1}^{\mathrm{n}} \lambda^{\mathrm{n-i}} \left[\frac{\mathrm{P}(s)}{1 + \mathrm{P}(s)\mathrm{G_c}(s)} \right]^2 \mathrm{e}'^2(\mathrm{i}) \tag{31}$$

The goal of the adaptive feedforward control algorithm is to find an optimal discrete filter and optimal weights so that the objective function of the cost function can be minimized, that is, the gradient of the cost function is zero:

$$\nabla_{\mathrm{k}}\xi = \frac{\partial \xi}{\partial \mathrm{w_k}} = -2\left[\frac{\mathrm{P}(s)}{1 + \mathrm{P}(s)\mathrm{G_c}(s)}\right]^2 \sum_{\mathrm{i}=1}^{\mathrm{N}} \{\lambda^{\mathrm{N-i}} \boldsymbol{y}(\mathrm{i}) [\mathrm{q}^*(\mathrm{i}) - \boldsymbol{w}^{\mathrm{T}}(\mathrm{i})\boldsymbol{y}(\mathrm{i})]\} = \boldsymbol{R}(\mathrm{n})\boldsymbol{w}(\mathrm{n}) - \boldsymbol{r}(\mathrm{n}) \tag{32}$$

$$\boldsymbol{R}(\mathrm{n}) = \sum_{\mathrm{i}=1}^{\mathrm{n}} \lambda^{\mathrm{n-i}} \boldsymbol{y}(\mathrm{i})\boldsymbol{y}^{\mathrm{T}}(\mathrm{i}) = \lambda \boldsymbol{R}(\mathrm{n}-1) + \boldsymbol{y}(\mathrm{n})\boldsymbol{y}^{\mathrm{T}}(\mathrm{n}) \tag{33}$$

$$\boldsymbol{r}(\mathrm{n}) = \sum_{\mathrm{i}=1}^{\mathrm{n}} \lambda^{\mathrm{n-i}} \boldsymbol{y}(\mathrm{i})\boldsymbol{q}^*(\mathrm{i}) = \lambda \boldsymbol{r}(\mathrm{n}-1) + \boldsymbol{y}(\mathrm{n})\mathrm{q}^*(\mathrm{n}) \tag{34}$$

where $\mathbf{R}(n)$ is the autocorrelation matrix of the interference signal, and $\mathbf{r}(\mathrm{n})$ is the cross-correlation matrix of the interference signal and the feedforward source signal. Then, the gradient of Eq. (32) is zero, and the arrangement can be obtained:

$$\hat{\boldsymbol{w}}(\mathrm{n}) = \boldsymbol{R}^{-1}(\mathrm{n})\boldsymbol{r}(\mathrm{n}) \tag{35}$$

For the convenience of description, define an inverse matrix $\mathbf{B}(n) = \mathbf{R}^{-1}(n)$, and the following expression can be obtained:

$$\begin{aligned}\boldsymbol{B}(\mathrm{n}) = \boldsymbol{R}^{-1}(\mathrm{n}) &= \lambda^{-1}\left[\boldsymbol{B}(\mathrm{n}-1) - \frac{\boldsymbol{B}(\mathrm{n}-1)\boldsymbol{y}(\mathrm{n})\boldsymbol{y}^{\mathrm{T}}(\mathrm{n})\boldsymbol{B}(\mathrm{n}-1)}{\lambda + \boldsymbol{y}^{\mathrm{T}}(\mathrm{n})\boldsymbol{B}(\mathrm{n}-1)\boldsymbol{y}(\mathrm{n})}\right] \\ &= \lambda^{-1}\left[\boldsymbol{B}(\mathrm{n}-1) - \boldsymbol{k}(\mathrm{n})\boldsymbol{y}^{\mathrm{T}}(\mathrm{n})\boldsymbol{B}(\mathrm{n}-1)\right]\end{aligned} \tag{36}$$

where $\mathbf{k}(n)$ is called the gain vector, and its expression is:

$$\boldsymbol{k}(\mathrm{n}) = \frac{\boldsymbol{B}(\mathrm{n}-1)\boldsymbol{y}(\mathrm{n})}{\lambda + \boldsymbol{y}^{\mathrm{T}}(\mathrm{n})\boldsymbol{B}(\mathrm{n}-1)\boldsymbol{y}(\mathrm{n})} \tag{37}$$

According to Eqs. (35)–(37), when the optimal solution of the tap weight vector has been obtained, the update formula of the weight vector can be further derived:

$$\hat{\boldsymbol{w}}(\mathrm{n}) = \boldsymbol{R}^{-1}(\mathrm{n})\boldsymbol{r}(\mathrm{n}) = \boldsymbol{B}(\mathrm{n})\boldsymbol{r}(\mathrm{n}) = \hat{\boldsymbol{w}}(\mathrm{n}-1) + \boldsymbol{k}(\mathrm{n})\left[\mathrm{q}^*(\mathrm{n}) - \boldsymbol{y}^{\mathrm{T}}(\mathrm{n})\hat{\boldsymbol{w}}(\mathrm{n}-1)\right] \tag{38}$$

Then, the initialization of the AHC algorithm is set to $\hat{\mathbf{w}}(0) = \mathbf{0}$, $\mathbf{B}(0) = \delta^{-1}$ and δ be very small positive numbers. The iterative formula of the AHC algorithm can be sorted out:

$$\begin{cases}\boldsymbol{k}(\mathrm{n}) = \dfrac{\boldsymbol{B}(\mathrm{n}-1)\boldsymbol{y}(\mathrm{n})}{\lambda + \boldsymbol{y}^{\mathrm{T}}(\mathrm{n})\boldsymbol{B}(\mathrm{n}-1)\boldsymbol{y}(\mathrm{n})} \\ \boldsymbol{B}(\mathrm{n}) = \lambda^{-1}\left[\boldsymbol{B}(\mathrm{n}-1) - \boldsymbol{k}(\mathrm{n})\boldsymbol{y}^{\mathrm{T}}(\mathrm{n})\boldsymbol{B}(\mathrm{n}-1)\right] \\ \hat{\boldsymbol{w}}(\mathrm{n}) = \hat{\boldsymbol{w}}(\mathrm{n}-1) + \boldsymbol{k}(\mathrm{n})\left[\mathrm{q}^*(\mathrm{n}) - \boldsymbol{y}^{\mathrm{T}}(\mathrm{n})\hat{\boldsymbol{w}}(\mathrm{n}-1)\right]\end{cases} \tag{39}$$

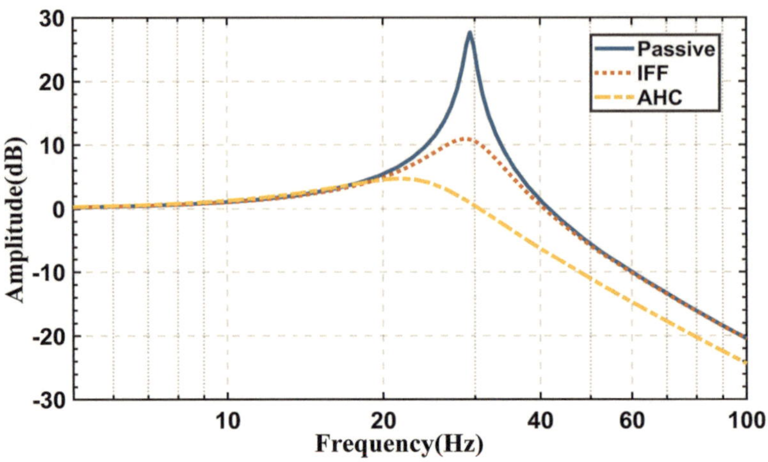

Figure 13.
Simulation comparison curves of transmissibility under different control modes.

The simulation analysis of the AHC is carried out in Matlab, and the transmissibility curve in **Figure 13** can be obtained. In passive control, the vibration of the load platform is not effectively suppressed at all, and it is significantly attenuated at high frequencies. In the IFF control, the resonance peak at the natural frequency of the system is obviously suppressed, but the high-frequency attenuation is not improved. When the AHC is adopted, the formant of the system is further reduced, and the high frequency also shows a higher attenuation. The piezoelectric AHC has a better vibration isolation effect.

5. Experimental verification of piezoelectric active vibration suppression

5.1 Piezoelectric active vibration suppression system construction

Through the above theoretical analysis, to verify the vibration suppression performance of the proposed AHC algorithm in the piezoelectric vibration suppression system, an experimental test system is built, as shown in **Figure 14**. **Figure 14a** shows a single-degree-of-freedom piezoelectric vibration suppression platform, which mainly includes a piezoelectric actuator and a passive suppression unit. The piezoelectric actuator is responsible for active suppression, and the passive vibration isolation unit provides system stiffness. The two ends of the vibration suppression platform are respectively provided with a load mass block and a basic mass block, and the basic mass block relates to the output shaft of the vibration exciter so that the suppression platform can receive external excitation. The whole vibration suppression platform is suspended horizontally by hanging ropes, which can ensure the free boundary conditions and introduce the influence of low stiffness and damping.

Figure 14b shows a photograph of the experimental equipment setup integration. The experimental system is mainly divided into a real-time active control system and spectrum test and analysis system. The real-time active control system consists of a charge amplifier, an NI controller, and a piezoelectric driver. The active control process is as follows: The charge amplifier amplifies the feedback signal of the force sensor into

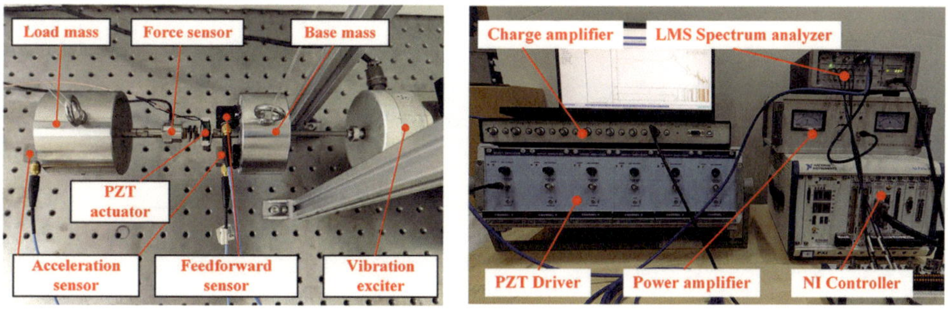

(a) (b)

Figure 14.
Piezoelectric vibration suppression system: (a) single-degree-of-freedom piezoelectric vibration sup-pression platform; (b) experimental equipment.

a voltage signal and outputs it to the NI controller. After the real-time active control algorithm in the NI controller, the control signal is generated and output to the piezoelectric driver for real-time active control of the linear piezoelectric actuator.

The spectrum test and analysis system include an LMS spectrum analyzer, an excitation signal output unit and an exciter power amplifier. The spectrum testing and analysis process are as follows: The LMS spectrum analyzer can output the excitation signal that simulates micro-vibration through the built-in simulation signal generator, and simulate the micro-vibration environment of the exciter through the exciter power amplifier. The LMS spectrum analyzer collects the vibration signals of the load platform and the base platform respectively through the acceleration sensor and performs postprocessing and spectrum analysis.

5.2 Verification of piezoelectric active vibration suppression performance

After completing the construction of the experimental system, the active control experiment was carried out in the piezoelectric suppression system. First, the system transmissibility curve is measured. The transmissibility curve measures the vibration transfer characteristics from the perspective of the frequency domain, which is a very important criterion. **Figure 15** shows the comparison curves of the system's transmissibility under different control methods. The blue line is the open-loop transmissibility curve of the system, which is not actively controlled and adopts pure passive vibration suppression. In the passive situation, the system has a peak near 30 Hz, which is the resonance peak, which means that the vibration amplitude will increase sharply here, and the vibration suppression effect is poor. At high frequencies, the passive transmissibility curve decays rapidly, which can be considered to have a suppressive effect. When IFF is used for piezoelectric active control, the resonance peak of the system at the natural frequency is attenuated to a certain extent, but the overall attenuation is not large, especially at high frequencies, the attenuation performance cannot be improved. When the AHC is used, the natural frequency of the system is moved forward, the resonance peak at the original natural frequency is greatly attenuated, and better attenuation is also reflected at high frequencies, which can effectively broaden the vibration isolation bandwidth. The specific values of the transmissibility curves under different control modes are shown in **Table 2**.

	Passive	IFF	AHC
Natural Frequency (Hz)	30	30	15
Resonance Peak (dB)	28	14.9	7.2
Decrease of Resonance Peak (dB)	/	13.1	20.8
Amplitude of 30 Hz (dB)	28	14.9	−1.9
Decrease of Amplitude in 30 Hz (dB)	/	13.1	29.9
Initial attenuation frequency (Hz)	43	43	22

Table 2.
The specific value of the experimental transmissibility curve.

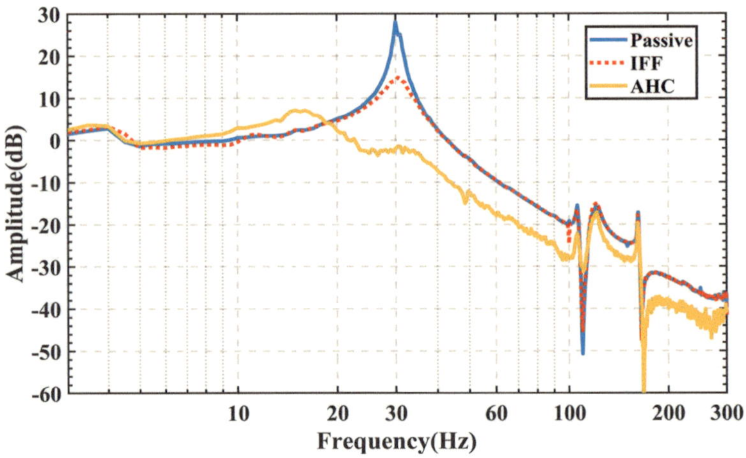

Figure 15.
Transmissibility curves under passive, IFF, and AHC.

The time-domain vibration signal can reflect the control effect more intuitively, but more details in the frequency domain cannot be revealed. **Figure 16a** shows the time domain acceleration signal when the piezoelectric vibration suppression system is without active control. The red line is the basic excitation signal, the blue line is the load response signal, and the load response signal is the final control target. It can be found that the piezoelectric system can also achieve a certain vibration suppression effect through passive vibration suppression, but it is generally difficult to meet the application requirements. **Figure 16b** shows the time domain information of the piezoelectric vibration suppression system under the AHC. Compared with the pure passive vibration suppression, after the AHC is turned on, the vibration signal at the load end is greatly attenuated. In the piezoelectric vibration suppression system, the AHC method can greatly improve the vibration suppression performance.

Figure 17 shows the self-power spectrum (PSD) curves with different control methods. In the case of passive vibration suppression (without control), the vibration energy of the load platform at the natural frequency is increased, which means that the vibration signal at the natural frequency is amplified. When AHC is used, the energy around the natural frequency is significantly suppressed because the piezoelectric actuator dissipates part of the energy.

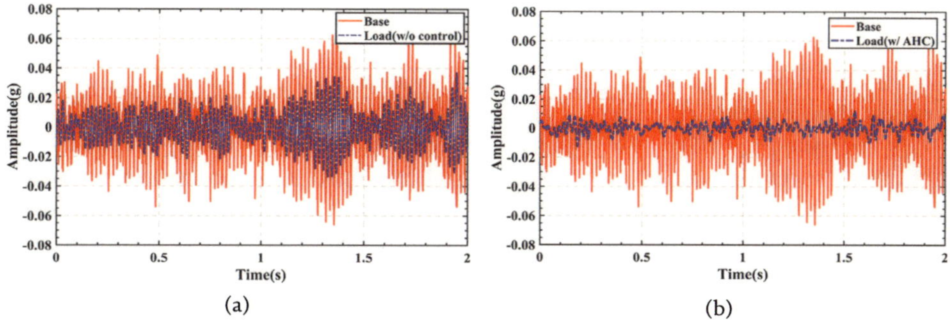

Figure 16.
Time domain amplitude comparison curve: (a) without AHC; (b) with AHC.

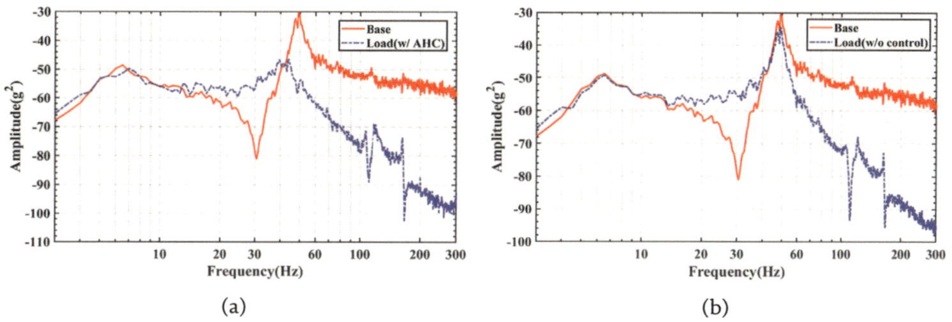

Figure 17.
Self-power spectral density comparison curve: (a) without AHC; (b) with AHC.

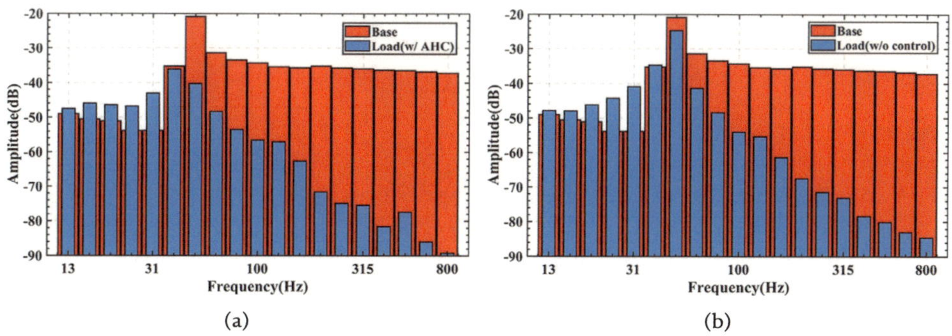

Figure 18.
One-third octave comparison curve: (a) without AHC; (b) with AHC.

Figure 18 shows one-third of octave curves with different control methods. When the piezoelectric vibration suppression system adopts passive vibration suppression (without control), the signal of the load platform near the natural frequency is significantly improved, which is also consistent with the self-power spectrum in **Figure 17**. When the AHC is turned on, the signal of the load platform near the natural frequency is attenuated, and the energy is dissipated by the piezoelectric actuator. This also shows that the piezoelectric AHC has a better inhibitory effect on the signal at the natural frequency.

6. Conclusions

In this chapter, the active vibration suppression of an active-passive composite vibration suppression system based on piezoelectric actuators is studied. On the basis of fully analyzing the characteristics of piezoelectric actuator and displacement amplifying mechanism and the dynamic model of vibration suppression system, an active composite control strategy based on IFF and RLS adaptive feedforward for vibration suppression on the piezoelectric system is discussed. The experimental results show that using the active control method, the vibration suppression system based on piezoelectric actuator has not only lower natural frequency, wider active vibration suppression bandwidth, but also reduces the value of the resonance peak, and maintains the attenuation rate at high frequency. Therefore, piezoelectric actuators are a good choice for vibration suppression, especially active vibration suppression.

Acknowledgements

The work was supported by the National Natural Science Foundation of China [grant No. 61903242]; Shanghai Sailing Program [grant No. 19YF1416200].

Conflict of interest

The author(s) declared no potential conflicts of interest with respect to the research, authorship, and/or publication of this article.

Author details

Min Wang[1,2]*, Songquan Liao[1], Xuan Fang[1] and Shibo Fu[1]

1 School of Mechatronic Engineering and Automation, Shanghai University, Shanghai, China

2 Ministry of Education, Engineering Research Center of Unmanned Intelligent Marine Equipment, Shanghai, China

*Address all correspondence to: xmwangmin@shu.edu.cn

References

[1] Yan B, Brennan MJ, Elliott SJ, Ferguson NS. Active vibration isolation of a system with a distributed parameter isolator using absolute velocity feedback control. Journal of Sound and Vibration. 2010;**329**(10):1601-1614

[2] Zuo L, Slotine JJE, Nayfeh SA. Model reaching adaptive control for vibration isolation. IEEE Transactions on Control Systems Technology. 2005;**13**(4):611-617

[3] Burkan R, Ozguney OC, Ozbek C. Model reaching adaptive-robust control law for vibration isolation systems with parametric uncertainty. Journal of Vibroengineering. 2018;**20**(1):300-309

[4] Sang H, Yang C, Liu F, Yun J, Jin G. A fuzzy neural network sliding mode controller for vibration suppression in robotically assisted minimally invasive surgery. International Journal of Medical Robotics and Computer Assisted Surgery. 2016;**12**(4):670-679

[5] Yang L, Lou J, Zhu S. Researches on magnetostrictive hybrid vibration isolation system based on sliding mode algorithm. In: Wen Z, Li T, editors. Foundations of Intelligent Systems (iske 2013) [Internet]. Berlin: Springer-Verlag Berlin; 2014. pp. 1095-1106. Available from: https://www.webofscience.com/wos/alldb/summary/1ba7f755-0d14-4def-ada3-274c2d75275f-201be7ef/relevance/1 [cited 2022 Jan 22]

[6] Wei K, Meng G, Zhang W, Zhou S. Vibration characteristics of rotating sandwich beams filled with electrorheological fluids. Journal of Intelligent Material Systems and Structures. 2007;**18**(11):1165-1173

[7] Manoharan R, Vasudevan R, Jeevanantham AK. Dynamic characterization of a laminated composite magnetorheological fluid sandwich plate. Smart Materials and Structures. 2014;**23**(2):025022

[8] Damanpack AR, Bodaghi M, Aghdam MM, Shakeni M. On the vibration control capability of shape memory alloy composite beams. Composite Structures. 2014;**110**:325-334

[9] Ebrahimi MR, Moeinfar A, Shakeri M. Nonlinear free vibration of hybrid composite moving beams embedded with shape memory alloy fibers. International Journal of Structural Stability and Dynamics. 2016;**16**(7):1550032

[10] Xu XS, Sun FM, Wang GP. The control and optimization Design of the Fish-like Underwater Robot with the aid of the Giant Magnetostrictive material actuator. Journal of Vibration and Control. 2009;**15**(10):1443-1462

[11] Pan Y, Zhao X. Design and analysis of a GMM actuator for active vibration isolation. In: 2015 IEEE International Conference on Mechatronics and Automation [Internet]. New York: IEEE; 2015. pp. 357-361. Available from: https://www.webofscience.com/wos/alldb/summary/8e1c2d38-d225-4850-8215-81e5f0e84b48-2012e459/relevance/1 [cited 2022 Jan 22]

[12] Manjunath TC, Bandyopadhyay B. Vibration control of Timoshenko smart structures using multirate output feedback based discrete sliding mode control for SISO systems. Journal of Sound and Vibration. 2009;**326**(1–2):50-74

[13] Wang C, Xie X, Chen Y, Zhang Z. Investigation on active vibration isolation of a Stewart platform with piezoelectric actuators. Journal of Sound and Vibration. 2016;**383**:1-19

[14] Jiang X, Zhu Y. Mechanical amplifier for Giant Magnetostrictive materials and piezoelectric materials [J]. Hydromechatronics Engineering. 2013

[15] Chen J, Zhang C, Xu M, Zi Y, Zhang X. Rhombic micro-displacement amplifier for piezoelectric actuator and its linear and hybrid model. Mechanical Systems and Signal Processing (MSSP). 2015;**50–51**:580-593

[16] Choi K-B, Lee JJ, Kim GH, Lim HJ, Kwon SG. Amplification ratio analysis of a bridge-type mechanical amplification mechanism based on a fully compliant model. Mechanism and Machine Theory. 2018;**121**:355-372

[17] Nguyen P-B, Choi S-B. A novel rate-independent hysteresis model of a piezostack actuator using the congruency property. Smart Materials and Structures. 2011;**20**(5):055003

[18] Wang DH, Zhu W, Yang Q. Linearization of stack piezoelectric ceramic actuators based on Bouc-wen model. Journal of Intelligent Material Systems and Structures. 2011;**22**(5): 401-413

[19] Badel A, Le Breton R, Formosa F, Hanene S, Lottin J. Precise positioning and active vibration isolation using piezoelectric actuator with hysteresis compensation. Journal of Intelligent Material Systems and Structures. 2014; **25**(2):155-163

[20] Minggang G, Zhi Q, Yanlong L. Sliding mode control with perturbation estimation and hysteresis compensator based on Bouc-wen model in tackling fast-varying sinusoidal position control of a piezoelectric actuator. Journal of Systems Science & Complexity. 2016; **29**(02):367-381

[21] Cheng L, Liu W, Yang C, Huang T, Hou Z-G, Tan M. A neural-network-based controller for piezoelectric-actuated stick-slip devices. IEEE Transactions on Industrial Electronics. 2018;**65**(3):2598-2607

[22] Nakamura Y, Nakayama M, Yasuda M, Fujita T. Development of active six-degrees-of-freedom micro-vibration control system using hybrid actuators comprising air actuators and giant magnetostrictive actuators. Smart Materials and Structures. 2006;**15**(4): 1133-1142

[23] Nguyen V-Q, Choi S-M, Choi S-B, Moon S-J. Sliding mode control of a vibrating system using a hybrid active mount. Proceedings of the Institution of Mechanical Engineers Part C Journal of Engineering Mechanical Engineering Science. 2009;**223**(6):1327-1337

[24] Jang D-D, Jung H-J, Shin Y-H, Moon S-J, Moon Y-J, Oh J. Feasibility study on a hybrid mount system with air springs and piezo-stack actuators for microvibration control. Journal of Intelligent Material Systems and Structures. 2012;**23**(5):515-526

[25] McInroy JE. Modeling and design of flexure jointed Stewart platforms for control purposes. IEEE/ASME Transactions on Mechatronics. 2002; **7**(1):95-99

[26] Preumont A, Horodinca M, Romanescu I, de Marneffe B, Avraam M, Deraemaeker A, et al. A six-axis single-stage active vibration isolator based on Stewart platform. Journal of Sound and Vibration. 2007;**300**(3–5):644-661

[27] Yang X, Wu H, Chen B, Kang S, Cheng S. Dynamic modeling and decoupled control of a flexible Stewart platform for vibration isolation. Journal of Sound and Vibration. 2019;**439**: 398-412

Chapter 8

A Review of Modeling and Control of Piezoelectric Stick-Slip Actuators

Zhenguo Zhang, Piao Fan, Yikun Dong, Shuai Yu, Keping Liu and Xiaohui Lu

Abstract

Piezoelectric stick-slip actuators with high precision, large actuating force, and high displacement resolution are currently widely used in the field of high-precision micro-nano processing and manufacturing. However, the non-negligible, non-linear factors and complexity of their characteristics make its modeling and control quite difficult and affect the positioning accuracy and stability of the system. To obtain higher positioning accuracy and efficiency, modeling and control of piezoelectric stick-slip actuators are meaningful and necessary. Firstly, according to the working principle of stick-slip drive, this paper introduces the sub-models with different characteristics, such as hysteresis, dynamics, and friction, and presents the comprehensive modeling representative piezoelectric stick-slip actuators. Next, the control approaches suggested by different scholars are also summarized. Appropriate control strategies are adopted to reduce its tracking error and position error in response to the influence of various factors. Lastly, future research and application prospects in modeling and control are pointed out.

Keywords: piezoelectric stick-slip actuators, dynamical model, friction model, conventional control, intelligent control

1. Introduction

With the rapid development of nanoscience and technology, precise positioning platforms and related technologies with micron motion range and nanoscale positioning accuracy have been widely applied in various fields, such as micromechanical systems, atomic force microscopy, and biomedical science [1–3]. In nanopositioning technology, system processing and measurement accuracy are increasingly demanding, so the resolution of nanopositioning needs to be achieved for the positioning system. Piezoelectric actuators are widely used in devices to achieve precise positioning because of their small sizes, simple structure, and high resolution. However, it cannot achieve long-distance motion owing to its limited motion range. Therefore, infinite stroke motion can be achieved by combining the piezoelectric actuator principle and the frictional inertia principle. The piezoelectric stick-slip actuator with high resolution and long stroke has advantages that other types of actuators do not have. The introduction and review of other types of actuators can be discovered in the literature [4]. The development and research of piezoelectric rod-slip actuators have been a very active field for several decades. A large number of studies on precision positioning stages based on rod-slip actuators have been carried out at home and abroad.

In the process of the stick-slip actuator, it is often affected by a variety of factors, such as the electrical and hysteresis characteristics of the piezoelectric element, dynamic characteristics of the piezoelectric stack, the mechanical structure, and friction characteristics between slider and drive rod. By analyzing the motion law of the main influencing factors, some non-significant nonlinear factors can be appropriately simplified in the process of establishing the model. Peng et al. proposed an end-effector model considering the linear dynamics and hysteresis characteristics of the piezoelectric stick-slip actuator, as well as the friction of the end-effector. However, the model is suitable for piezoelectric stick-slip actuators with horizontal movement of the end-effector. The effect of gravity should be taken into account in subsequent model studies [5]. The structure of the piezoelectric stick-slip actuator is simple, but since the system relies on the frictional drive for long-distance positioning, there are uncertainties and disturbances. To achieve great motion characteristics, the control requirements of the system are extremely important. The control method is mainly divided into open-loop control and closed-loop control. The open-loop control is often used to compensate for the influence of hysteresis characteristics, and the adjustment and control method is used to realize positioning and tracking. Song et al. proposed feedforward control based on the Preisach model to eliminate the hysteresis characteristics [6]. Li et al. proposed a parasitic-type piezoelectric actuator, and this control strategy was used in scanning mode (high precision motion) to improve the positioning accuracy simply and effectively [7].

In the past few decades, a lot of research has been conducted on the modeling and control of hysteresis nonlinearity of piezoelectric actuators. In this paper, we plan to focus on the piezoelectric stick-slip actuator. Section 2 introduces various models and modeling methods under different characteristics of piezoelectric stick-slip actuators and summarizes the overall comprehensive model including each sub-model. Section 3 reviews various control schemes of the piezoelectric stick-slip actuator, including open-loop control and closed-loop control, and presents its limitations and related challenges. Finally, summarizes the problems in modeling and control of the piezoelectric stick-slip actuator and the future development direction in Section 4.

2. Modelling of piezoelectric stick-slip actuators

The piezoelectric stick-slip actuator is a type of actuator that uses the inverse piezoelectric effect and the friction principle to realize the stepping displacement. Its motion process is more complex and influenced by more factors. Therefore, the establishment of a piezoelectric stick-slip actuator model needs to reflect as many motion laws as possible. By analyzing the principle of stick-slip drive and establishing a comprehensive representative model of piezoelectric stick-slip drive based on a simplified model. On the one hand, the influence of each factor on the system can be analyzed by simulation. On the other hand, for the control method containing the system model, the establishment of the model is the theoretical basis for control, and the accuracy of its model directly affects the performance of the control system.

A typical piezoelectric stack-frictional rod type consists of four principal parts—the fixed part, the piezoelectric stack, the frictional rod, and the slider. The piezoelectric element elongates and contracts under the action of a sawtooth wave drive signal driving the friction rod into a corresponding reciprocating motion. The slider is displaced forward by the frictional force with the friction rod and its own inertia. The basic drive principle of a piezoelectric stick-slip actuator is illustrated in **Figure 1**. The drive process is divided into a stick phase and a slip phase within one cycle, and its force analysis is shown in **Figure 1b**. In the sticky stage, when a slowly increasing voltage is added to the piezoelectric element, the piezoelectric element slowly extends to drive the friction rod to the right. At this time, the friction force between the slider and the friction rod is greater than the inertia force. The slider and the friction rod remain relatively stationary and move together with the right S_0. At the slip stage, the voltage loaded on the piezoelectric element disappears quickly, and the piezoelectric element contracts quickly to the initial position. At this time, the inertia force between the slider and the friction rod is more important than the friction force and drives the slider to produce a backward displacement S_1 to the left. The effective step in each cycle is ΔS. The drive can achieve continuous motion to the right by continuously repeating the motion process.

Research shows that the modeling accuracy of piezoelectric stick-slip actuators is mainly determined by the following aspects—the hysteresis effect of the piezoelectric stack, the relationship between the driving voltage and the driving force, the model of the mechanical structure of the piezoelectric stick-slip actuator, and the friction model between the slider and the friction rod. Therefore, a representative integrated model of piezoelectric stick-slip actuators is discussed based on these factors in this paper.

2.1 Electromechanical coupling model for piezoelectric stick-slip actuators

The piezoelectric actuator part is an electromechanical coupling system. When a certain drive voltage is applied, the piezoelectric actuator generates a certain displacement and output force because of the inverse piezoelectric effect. The modeling of the piezoelectric actuator needs to reflect the relationship between the driving

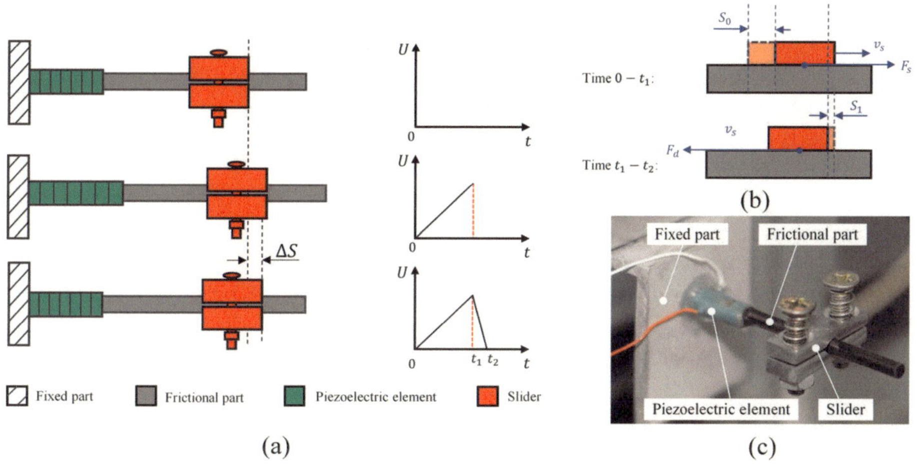

Figure 1.
Operation principle. (a) Driving principle of the stick-slip actuator. (b) Force analysis of the stick-slip actuator. (c) The actual object of the stick-slip actuator.

voltage and the deformation and output force generated by the piezoelectric actuator at the same time, which are coupled with each other.

To quantitatively analyze a system, it is necessary to describe the dynamics of the system through a mathematical model. This enables more information about the system to be described, resulting in better system control. Adriaens et al. pointed out that if the piezoelectric positioning system is properly designed, a second-order approximate modeling approach can be used to represent the dynamics of the system very well [8]. Typically, it can be viewed as a simplified spring-damped mass second-order system with a friction bar. Its linear dynamics is represented as

$$m\ddot{x} + c\dot{x} + kx = F_p - F_f \tag{1}$$

where x is the displacement of the slider, c and k denote the damping and stiffness of the piezoelectric drive stack, and m denotes the total mass of the piezoelectric stick-slip actuator. F_p is the output force of the piezoelectric stack, and F_f is the frictional reaction force between the slider and the friction rod.

There are both hysteresis and creep effects in the process of applying a voltage to the ends of the piezoelectric stack. When the nonlinear characteristics of the system are not considered, the piezoelectric actuator input voltage and output force can be expressed as

$$F_p = K_h u(t) \tag{2}$$

where K_h is the conversion ratio of input voltage to output force and $u(t)$ is the drive voltage of the upper piezoelectric driver.

2.2 Hysteresis model for piezoelectric stick-slip actuators

In the ideal case, the output displacement of the piezoelectric stick-slip actuators is linearly related to the input control voltage curve. Due to the inherent characteristics of this material, there are certain nonlinear characteristics such as the hysteresis effect and creep phenomenon in the actual driving process. However, because the creep effect is so small that it is mostly ignored and the hysteresis effect is mainly considered in the existing literature. The hysteresis phenomenon refers to the non-coincidence between the boost displacement curve and the bulk displacement curve when the drive voltage is applied to the piezoelectric driver. The hysteresis model is used to represent the force generated by the piezoelectric driver with a new equation as

$$m\ddot{x} + c\dot{x} + kx = H(t) - F_f \tag{3}$$

Currently, they can be broadly classified into physical and phenomenological models based on the modeling principles. Physical models are based on the physical properties of materials, among which the Jiles-Atherton model and Ikuta-K model are more common [9, 10]. However, physical modeling is based on the physical properties of the material, so the model implementation is difficult and affects the generality of the physical hysteresis model. Phenomenological models are based on input-output relations of hysteresis systems and are described using similar mathematical models. There are three broad categories based on the modeling approach, operator hysteresis models, differential equation hysteresis models, and intelligent hysteresis models.

2.2.1 Operator hysteresis model

The common operator hysteresis models are the Preisach model, Prandtl-Ishlinskii (PI) model, and Krasnosel'skii-Pokrovskii (KP) model.

The Preisach model was first used to describe the hysteresis phenomenon in ferromagnetic materials. It was then gradually extended to describe the hysteresis behavior of smart materials, such as piezoelectric ceramics and magnetically controlled shape memory alloys. It is one of the most frequently studied nonlinear models of hysteresis [11]. The model consists of an integral accumulation of relay operators in the Preisach plane. The relay operator is shown in **Figure 2a**. Its mathematical expression is given by

$$y(t) = \iint_P \mu(\alpha, \beta) \gamma_{\alpha\beta}[u(t)] d\alpha d\beta \tag{4}$$

where $u(t)$ and $y(t)$ represent the input and output of the Preisach model, $\mu(\alpha, \beta)$ is the weight function of the relay operator corresponding to the Preisach plane, $\gamma_{\alpha\beta}$ represents the output of the relay operator, α and β are the switching thresholds of the relay operator.

Based on the previous work, Li et al. proposed that a multilayer neural network can be used to approximate the Preisach model. It can use any algorithm trained for neural networks to identify the model. The model is more flexible to adapt to different working conditions than the conventional model [12]. Later, Li et al. proposed a transformation operator for neural networks that can transform the multi-valued mapping of Lagrange into a one-to-one mapping. By adjusting its weights, the neural network model is made applicable to different operating conditions. The drawback that the Preisach model cannot be updated online is solved [13].

Both the PI model and the KP model evolved from the Preisach model. The PI model has a single threshold and two continuum hysteresis operators, the reciprocal inverse play operator and the stop operator [14], as shown in **Figure 2b** and **2c**. Therefore, the PI model can be derived from the PI inverse model by the stop operator, which can be easily used to design feedforward control compensators by the inverse model. The KP model uses a modified and improved Play operator with its corresponding density function superimposed for hysteresis modeling. While the traditional play operator has only one threshold that determines its width and

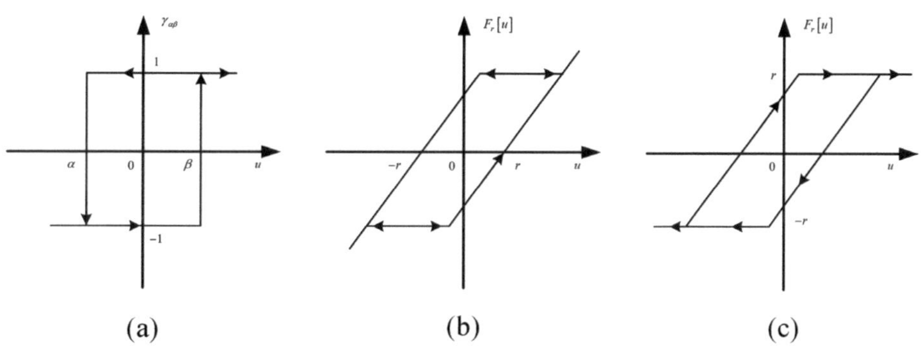

Figure 2.
Operators. (a) Relay operator. (b) Play operator. (c) Stop operator.

symmetry, the KP operator has two different thresholds, enabling it to describe more complex hysteresis nonlinear behavior [15].

2.2.2 Differential equation hysteresis model

The common differential equation hysteresis models include the Duhem hysteresis model, the Bouc-Wen hysteresis model, and the Backlash-like hysteresis model. P. Duhem et al. proposed the Duhem model, which is a differential equation. The model was later improved by Coleman and Hodgdon and applied to describe the hysteresis behavior of piezoelectric ceramics [16]. Its common mathematical expression is given by

$$\dot{x} = \alpha_D |\dot{u}| [f(u) - x] + g(u)\dot{u} \tag{5}$$

where x and u represent the output displacement and input voltage of the Duhem model, and α_D represents the model parameters of the Duhem model, which is a positive constant. The functions $f(u)$ and $g(u)$ determine the shape and performance of the input-output hysteresis curve of the Duhem model.

Although the Duhem model is also applied to describe the piezoelectric ceramic hysteresis problem, its application in engineering is greatly limited due to the difficulty of solving the model inverse model. Su et al. proposed a simplified dynamic hysteresis Backlash-like model based on the Duhem model [17]. This model has fewer parameters compared to the Duhem model and is a first-order differential equation with the mathematical expression

$$\dot{x} = \alpha_B |\dot{u}| [cu - x] + \beta_B \dot{u} \tag{6}$$

where x and u represent the output displacement and input voltage of the Backlash-like model, α_B, c and β_B are constants.

The Bouc-Wen model was originally proposed as a differential equation by Bouc [18] and was later refined by Wen [19] to form the current Bouc-Wen model. The classical Bouc-Wen model can describe a large class of hysteresis phenomena and has a concise expression [20], which is given as follows

$$\begin{cases} y(t) = kv(t) - h(t) \\ \dot{h}(t) = \alpha\dot{v}(t) - \beta|\dot{v}(t)||h(t)|^{n-1}h(t) - \gamma\dot{v}(t)|h(t)|^n \end{cases} \tag{7}$$

where k denotes the scale factor of the system input to the output of the hysteresis part, α, β, γ , and n denote the parameters of the hysteresis part of the model. The Bouc-Wen output $y(t)$ consists of a proportional linear part $kv(t)$ and a hysteresis nonlinear part $h(t)$.

The Bouc-Wen model is simple in form and has few identification parameters, which is convenient for controller design. However, it cannot completely describe the hysteresis characteristics of piezoelectric ceramics, its accuracy is low and it is only applicable to single frequency signals. The Bouc-Wen model is difficult to accurately describe the hysteresis phenomenon under the effect of frequency signals [21]. Therefore, the application of this model in practical engineering is greatly limited.

2.2.3 Intelligent hysteresis model

In addition to the above hysteresis models, there are some other classes of hysteresis models used in hysteresis modeling of smart materials, such as neural network models polynomial models, and other nonlinear models. Gan et al. proposed a polynomial model for the hysteretic nonlinearity of piezoelectric actuators. Experimental results show that the proposed model has higher modeling accuracy than the conventional PI model [22]. Cheng et al. proposed a method for nonlinear model prediction. First, a multilayer neuron network is used to identify the nonlinear autoregressive sliding average model of piezoelectric ceramics. Then, the tracking control problem is transformed into an optimization problem for model prediction. Finally, the Levenberg-Marquardt method is used to solve the numerical solution of the nonlinear minimization [23].

There are some other models, for example, Zhang et al. proposed a proposed Rayleigh model to describe the hysteresis characteristics of the piezoelectric drive system. The parameters of the rate-dependent Rayleigh model were obtained and validated based on the functional and experimental data [24]. Li et al. proposed a simplified interval type 2 (IT2) fuzzy system for hysteresis modeling of piezoelectric drives. In the experiments, gradient resolution and inverse resolution are used to identify the IT2 fuzzy hysteresis model [25]. Although these models are not as widely applied as the three major classes of models, they can often achieve good results in some cases when dealing with the hysteresis characteristics in some specific situations.

2.3 Selection of friction model

The output of the drive system is ultimately transferred to the slider in the form of friction, so the choice of friction model will directly affect the accuracy of the stick-slip drive platform model. At present, with the in-depth research of many international scholars on friction models, a variety of friction models have been established, which can be broadly divided into two categories, static friction models and dynamic friction models. Static friction models describe the friction force as a function of relative velocity. The dynamic friction model describes the friction force as a function of relative velocity and displacement. In contrast to static friction, which only considers the case where the relative velocity is not zero, the dynamic friction model uses differential equations to refer to the case where the relative velocity speed is zero. Therefore, in terms of accuracy, the dynamic friction model is more comprehensive and realistic than the static model. However, in addition to the accuracy of the friction model, it is also necessary to consider the complexity of the model, not all cases need to use the dynamic friction model.

2.3.1 Static friction model

The most widely used models in static friction modeling can be broadly classified into a series of coulomb and the stribeck model. Leonardo da Vinci took the lead in discovering that friction is related to the mass of an object and constructed a model. The model considers that the frictional force is proportional to the mass of the object and opposite to the direction of motion. Later this model was improved and called the coulomb model with the expression for friction

$$F_f = F_c \operatorname{sgn}(v) \tag{8}$$

where F_f is the friction force, F_c is the Coulomb friction force, and $\operatorname{sgn}(v)$ is the sign function.

In some studies, classical friction models have been used to represent the friction between the slider and the friction bar. The four static friction models commonly used in the early days are shown in **Figure 3** [26]. However, the slider step is only a few tens of nanometers to a few microns. The friction at this point is determined by the pre-slip displacement, which is the motion of the object before it is about to slide formally. When the friction surface is rough, the Coulomb friction model cannot accurately predict the friction force at pre-slip. Experiments have shown that in the pre-slip domain, the friction force depends on the micro-displacement between the two contacting surfaces [27]. However, the Coulomb friction model does not accurately predict this effect and will result in a relatively large error in the system. Therefore, a more accurate friction model is needed to describe the friction between the drive block and the terminal output.

These models need to exhibit some important static and dynamic properties of friction, such as the stribeck effect, Coulomb friction, stick friction, and pre-slip displacement. Li et al. proposed the stribeck model, which was the first model to describe dynamic and static frictional transition processes [28]. Its mathematical expression is as follows

$$F_f = \left(F_c + (F_s - F_c)e^{-|v/v_s|\varsigma_s}\right) \operatorname{sgn}(v) + bv \tag{9}$$

where F_s is the maximum static friction, F_c is the Coulomb friction force, b is the coefficient of stick friction, v_s is the stribeck effect velocity value, and ς_s is the empirical constant.

It not only reflects the linear relationship between dynamic friction and velocity but also expresses the change of friction during the transition between dynamic and static friction. It lays the groundwork for future research and the establishment of a dynamic friction model.

2.3.2 Dynamic friction models

Dahl et al. proposed the Dahl model, which is a dynamic friction model [29]. It describes for the first time the pre-slip displacement in a friction model, represented by the partial differential equation

$$\frac{dF_f}{dx} = \sigma\left(1 - \frac{F_f}{F_c}\operatorname{sgn}(v)\right)^{\alpha} \tag{10}$$

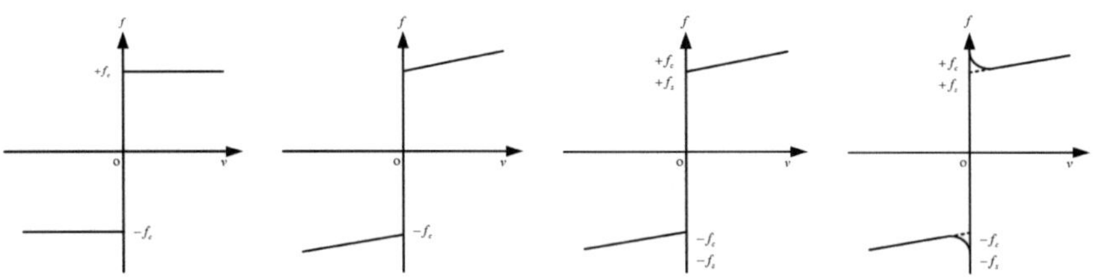

Figure 3.
Four classical static friction models. (a) Classical Coulomb friction model. (b) Coulomb and stick friction models. (c) Static friction, coulomb, and stick friction models. (d) Stribeck model.

where x is the shape variable and σ is the stiffness coefficient. α determines the shape of the curve.

However, the model does not capture the variation of the static friction phase and also does not explain the stribeck phenomenon. It is far from an adequate description of the stick-slip-driven friction interface. However, due to its simple and accurate expressiveness, it provides a solid foundation for the subsequent dynamic modeling.

Based on the study of the Dahl model, the French scholar Canudas proposed the LuGre model [30]. Its principle formula is

$$\begin{cases} F_f = \sigma_0 z + \sigma_1 \dfrac{dz}{dt} + \sigma_2 v \\ \dfrac{dz}{dt} = v - \dfrac{z}{g(v)}|v| \\ g(v) = \dfrac{1}{\sigma_0}\left[F_c + (F_s - F_c)e^{-(v/v_s)^2}\right] \end{cases} \tag{11}$$

where α_0 denotes the stiffness coefficient of the elastic bristle, α_1 denotes the system damping coefficient, α_2 is the coefficient of stick friction, v is the relative velocity of the object surface, z is the mean deformation of the friction surface, and $g(v)$ is the described stribeck effect.

The LuGre model introduced the stribeck effect in addition to combining the idea of pre-slip displacement in the Dahl model. The idea of the bristle effect was designed to address the changing situation of the static friction phase. Swevers et al. proposed a new structure of the dynamic friction model [31]. The non-local memory hysteresis function and the modeling of arbitrary transition curves were added on the basis of the LuGre model. This allows the model to accurately describe the experimentally obtained friction characteristics, stribeck friction during slip, hysteresis behavior during slip, and stick-slip behavior. Since the structure of the obtained model is flexible, it can be further extended and generalized.

2.4 Comprehensive model of piezoelectric stick-slip actuators

By combining the above models and considering other influencing factors, a comprehensive dynamics model considering the electrical model of the piezoelectric stick-slip actuator, the hysteresis effect, the linear dynamics performance, and the frictional characteristics of the system can be obtained, as shown in **Figure 4**.

In the stick-slip drive system, the output force of the piezoelectric stack is first obtained by the change of voltage at both ends of the piezoelectric element. Then, the electromechanical conversion model of the drive transmission system composed of the piezoelectric stack and the flexible transmission mechanism is transformed into the displacement and force output of the transmission system. Finally, the displacement is transferred to the slider by the kinematic friction conversion, and the final

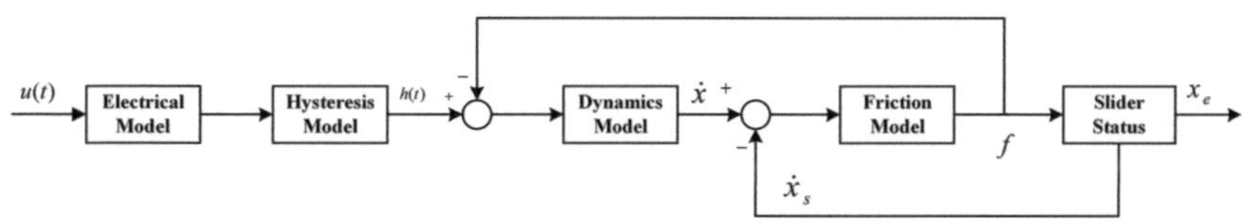

Figure 4.
Flowchart of a comprehensive model of a piezoelectric stick-slip actuator.

displacement output is obtained. The mathematical equation of its integrated model is as follows

$$\begin{cases} m\ddot{x} + c\dot{x} + kx = H(t) - F_f \\ m_s\ddot{x}_s = F_f \\ x_e = x - x_s \end{cases} \tag{12}$$

where $H(t)$ can be given by the previous hysteresis model [i.e., (4)–(7)]; F_f can be given by the friction model [i.e., (8)–(11)]; x_s is the backward displacement of the slider under the action of dynamic friction; and x_e is the forward displacement of the slider.

Wang et al. developed a kinetic friction model of the actuator and investigated the effect of the input drive voltage on the viscous slip motion of the actuator through simulation [32]. Nguyen et al. used the method of dimensionality reduction to describe the frictional contact behavior of the stick-slip microactuator. The model accurately predicts the frictional contact behavior of the actuator on different geometric scales without using any empirical parameters [33]. Piezoelectric stick-slip actuators also have more complex dynamic characteristics. The ability to simulate a wide range of dynamic characteristics is the direction of increasing research. Shao et al. found that the contact behavior of the piezoelectric feed element produced an inconsistent displacement response. So the Hunt-Crossley kinetic model, LuGre model, distributed parameter method combined with Bouc-Wen hysteresis model were used to model the viscous slip actuator. This model can effectively model the step inconsistency in the front-to-back direction of the actuator [34]. Wang et al. proposed a stick-slip piezoelectric actuator dynamics model considering the overall system deformation. The model introduced stiffness coefficients and damping coefficients for the whole system and successfully simulated three single-step characteristics, namely, backward motion, smooth motion, and a sudden jump for the first time [35]. Due to a large number of parameters in the dynamic model, the accurate identification of each parameter will be quite difficult. Therefore, more accurate identification of simulation parameters needs further research, which may be our future work.

3. Control schemes of piezoelectric stick-slip actuators

The piezoelectric stick-slip actuator introduces mechanical structures, such as friction rods and linear cross roller guides on the basis of piezoelectric stacks. A sawtooth wave signal is applied to the piezoelectric stack to achieve a stepping large stroke and high precision motion. In the actual application process, the traditional mechanical structure can no longer meet the demand of real positioning accuracy due to the complex nonlinear effects in the system and the influence of external disturbances. Therefore, intelligent control algorithms combined with computer hardware devices are usually introduced to eliminate or reduce the impact of the above problems on motion accuracy. This section mainly discusses the existing controller design methods from two parts—open-loop control and closed-loop control.

In this paper, control schemes and hardware facilities for piezoelectric stick-slip actuators are briefly described in most of the literature. Depending on whether a closed-loop is formed, the main categories are feedforward control, feedback control, and composite control with a combination of feedforward and feedback. The inverse model-based control is mostly feedforward control, which is usually used to compensate for the hysteresis characteristics of piezoelectric stick-slip actuators. The control system is shown in **Figure 5**, y_d is the expected input, v_{inv} is the theoretical input under the expected output value, y is the actual output value.

Feedback control is the real-time feedback of the actual measured displacement through data measurement equipment, such as sensors, and the feedback value is one of the inputs of the controller. Feedback control can effectively improve the robustness of the control system, and its control mode is shown in **Figure 6**. Where, v is the output of the feedback controller.

In actual control, simple feedforward control and feedback control cannot meet the control demand. Therefore, in most cases, researchers combined the advantages of feedforward control and feedback control, usually the feedforward control and feedback control of the compound control scheme are applied to the piezoelectric stick-slip actuator position tracking control. The control system is shown in **Figure 7**.

3.1 Conventional open-loop control of piezoelectric stick-slip actuators

Feedforward control is essentially an open-loop control. However, in the overall control of piezoelectric stick-slip actuators, the advantages of this control method are very limited. This has been found by many scholars—since the voltage changes for each step displacement are very fast, there is almost no hysteresis, creep effect, and the use of feed-forward control in this motion mode is not necessary.

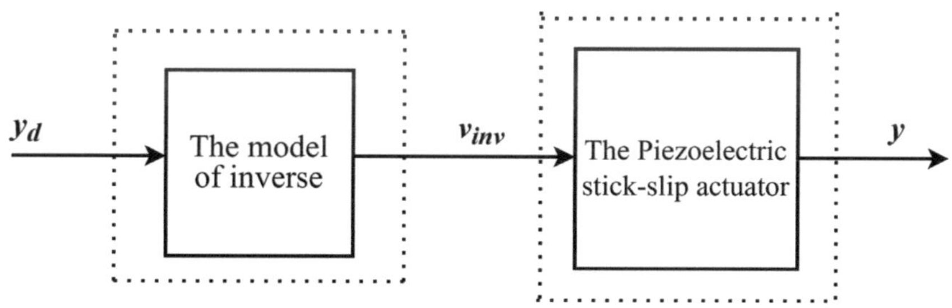

Figure 5.
Feedforward control system principle based on inverse model.

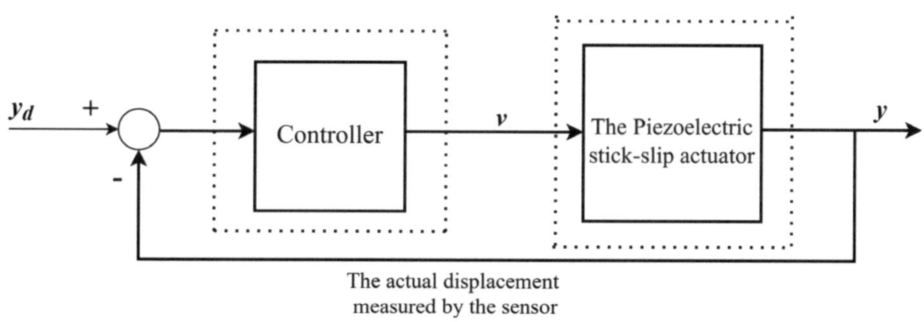

Figure 6.
Feedback control system principle.

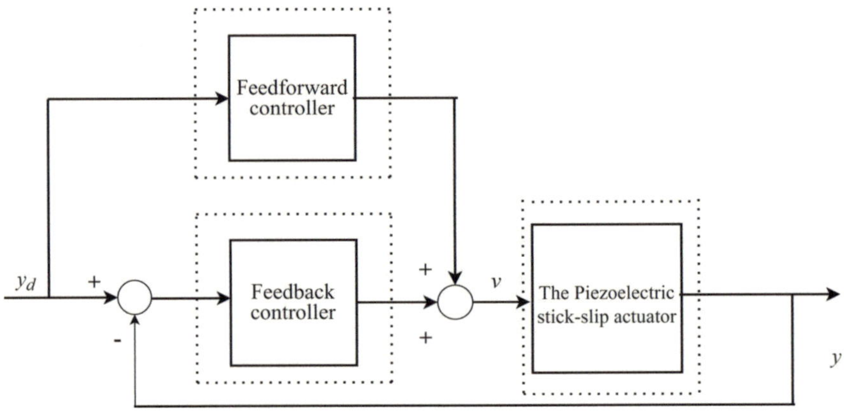

Figure 7.
Feedforward and feedback compound control system.

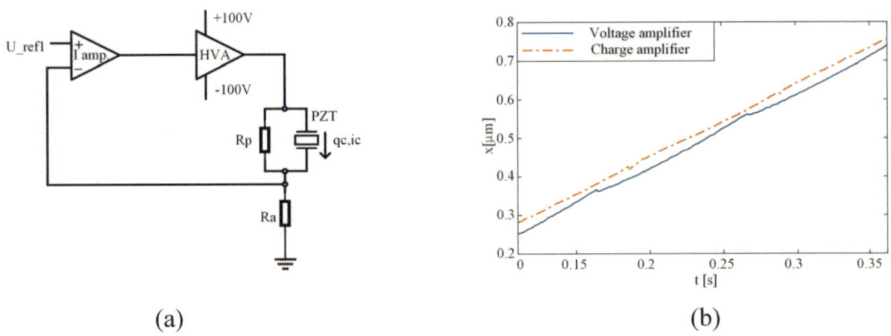

Figure 8.
(a) A simplified circuit diagram. (b) Experiment effect.

Feedforward control is often used to improve the quality of the end motion output of piezoelectric stick-slip actuators. Holub et al. compensated for the hysteresis of piezoelectricity by varying the amplitude of the input voltage for position errors and hysteresis modeling [36]. Chang et al. compensate for hysteresis by changing the phase lag of the actuator [37].

Feedforward control of end-effectors is challenging as the control accuracy of feedforward control is heavily dependent on the accuracy of the above model. The main source of difficulty comes from the many problems in the real system, including the hysteresis of the piezoelectric, the creeping nature, and the non-linearity of the frictional motion, the vibration between the stick and sliding points, the wear of the material between the movers and the stator, and other uncertainties [38].

Another feed-forward control by some scholars-charge control to compensate for hysteresis in the actuator. Špiller et al. developed a hybrid charge-controlled driver that slides by generating a high-voltage asymmetric sawtooth wave and feeding it into a capacitive load to compensate for the piezoelectric hysteresis as well as to achieve fast back-off. A simplified circuit diagram of the piezoelectric driver is shown in **Figure 8a**. This control method combines a charge control scheme with a switch is an effective solution, and the proposed hybrid amplifier has better motion linearity as shown in **Figure 8b** [39].

The control of piezoelectric stick-slip actuators is usually divided into one-step control stage and sub-step control stage, also known as step mode and scan mode. Step mode refers to the piezoelectric stick-slip actuator moving forward at a fixed step size and a fixed frequency when it is far from the desired position. Until the error between the actual position and the desired position is less than the single-step displacement of the piezoelectric stick-slip actuator, the precise positioning is realized by controlling the elongation of the piezoelectric stack, which is called the scanning control stage. In the stepping mode, the voltage changes rapidly and there is almost no creep. At the same time, in the stepping mode, the control accuracy is not necessary. Therefore, in the control process of stick-slip motion, there are few overall feedforward control cases. And feedforward control is usually implemented in the scanning control stage. The core component of the piezoelectric stick-slip actuator is the piezoelectric stack, which moves through the inverse piezoelectric effect of the piezoelectric stack. Compared with the motion control of the piezoelectric stick-slip actuator, feedforward motion control of the piezoelectric stack actuator is more mature. Chen of Harbin Institute of Technology defined a new function named mirror function, which connected the dynamic hysteresis model with the classical Preisach model and established a new dynamic hysteresis model to describe the input-output relationship of the piezoelectric actuator under different conditions. On this basis, a feedforward control scheme based on the dynamic hysteresis inverse model is designed [40].

In addition, Ha et al. experimentally identified the hysteretic parameters of the Bouc-Wen model, and on this basis designed a feedforward compensator to compensate for the influence caused by the nonlinearity of the hysteretic effect. Finally, the simulation results of the compensator and the designed voltage waveform are given to realize the feedforward control of the piezoelectric stack [41]. Wei et al. proposed a feedforward controller based on an improved rate-dependent PI hysteresis inverse model, which achieved the expected effect [42]. In recent years, Zhang et al. also proposed a third-order rate-dependent Rayleigh model to describe the hysteresis nonlinearity of piezoelectric stacks. And proposed a feedforward control scheme based on the inverse third-order rate-dependent Rayleigh model, which also verified the effectiveness of the method through experiments [43]. Feedforward control often plays an obvious role in hysteresis compensation. In the future piezoelectric stick-slip drive controller design, the existing piezoelectric stack feedforward control methods can be used for reference to realize the feedforward control in the scanning stage. Simple feedforward control has poor robustness in the application, so most researchers use compound control to improve the control accuracy.

3.2 Conventional closed-loop control of piezoelectric stick-slip actuators

Piezoelectric stick-slip actuators are affected in practice by factors such as environmental vibration and their own nonlinear characteristics, and their controllability becomes poor. Therefore, appropriate control methods are needed for closed-loop control to meet the actual working requirements. Zhong B et al. found that differences in object surface roughness and wear can cause inconsistent velocities during the movement of a piezoelectric stick-slip actuator. Therefore, a dual closed-loop controller for velocity and position was designed to achieve high accuracy positioning. Its principle of double closed-loop control is shown in **Figure 9**. The experimental results show that the standard deviation of the speed of the controller is less than 0.1 mm/s, and the repeated positioning accuracy reaches 80 nm, both of which achieve a good control effect [44]. The design of the controller considers a more accurate speed

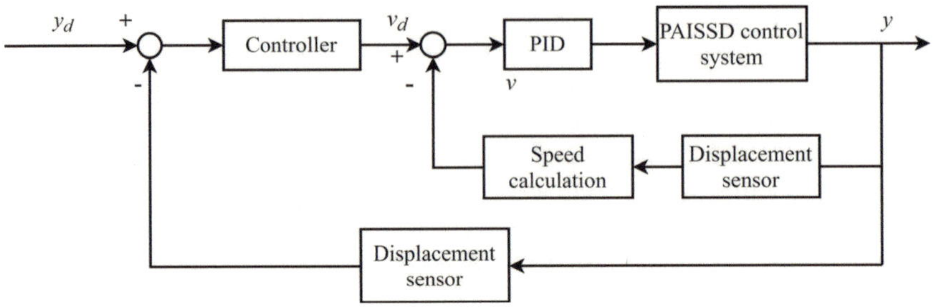

Figure 9.
Double closed-loop control principle.

control scheme, which has a high reference value for realizing the fast positioning of piezoelectric stick-slip actuators. Rong et al. introduced strain gauge as positioning sensor of the precision manipulator with piezoelectric stick-slip actuators and developed a displacement prediction method based on this. The feedforward PID control method is used throughout the system to improve the dynamic performance of the system. As shown in **Figure 10a**, it is the simulation of displacement-prediction control positioning performance (target displacement of 5.5 μm). **Figure 10b** shows the comparison of displacement response under the open-loop control and displacement response under the displacement-prediction control (target displacement of 20 μm). It can be seen from the experimental results that the 200 nm steady-state error of the proposed control method is much lower than that of the open-loop control [45]. The commonly used classical control methods are difficult to achieve high control accuracy in practical applications due to the limitations of parameters, weak automatic regulation and poor robustness.

Rakotondrabe et al. designed a micro-positioning device based on a stick-slip actuator. The control process is divided into a step mode and a scan mode, where the scan mode is precisely controlled by a PI controller [46]. Theik et al. used an inertial piezoelectric actuator to suppress the vibration of the hanging handle and designed three controllers—PID manual setting, PID self-setting, and PID-AFC. The best damping effect was achieved by experimentally comparing the PID-AFC controller. When the mass of the inertia block is larger, the vibration damping effect is more obvious [47]. These control methods are developed by the classical control theory in the actual control process of piezoelectric stick-slip actuators.

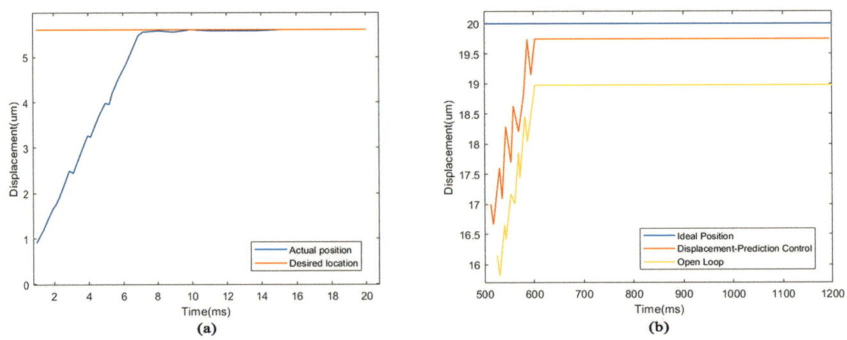

Figure 10.
(a) Simulation of displacement-prediction control positioning performance (target displacement of 20 μm). (b) Comparison between displacement response under the open-loop control and displacement response under the displacement-prediction control (target displacement of 20 μm).

3.3 Intelligent control of piezoelectric stick-slip actuators

By introducing intelligent control algorithms, such as sliding mode control algorithm and neural network algorithm, the self-adjusting control of piezoelectric stick-slip actuator is realized. Closed-loop control with feedback is a common control mode of piezoelectric stick-slip actuators, which can effectively compensate for the effects of hysteresis nonlinearity, complex friction relations and external interference on the positioning accuracy, and improve the robustness of the controller. The closed-loop control is mainly divided into two kinds of closed-loop control. The first closed-loop control is the voltage amplitude of the driving signal, which adjusts the single-step size of the piezoelectric stick-slip actuator by controlling the voltage amplitude. The other is the control of the driving signal frequency. By adjusting the frequency of the piezoelectric stick-slip actuator, the speed of the piezoelectric stick-slip actuator can be controlled. Cao et al. proposed a sliding mode control method based on linear autoregressive proportional integral-differential. It can solve the problem that the hysteretic characteristics of piezoelectric stacks in piezoelectric stick-slip actuators and the nonlinear friction relationship between end-effector and workbench affect the control effect. Firstly, an ARX model of the system is designed, and its state space description is obtained. Then, the sliding mode control is introduced, and PID control is introduced as the frequency switching controller in the sliding mode control, so that the error tends to zero, to achieve better speed control [48].

In addition to introducing the inherent mathematical model into the controller, the controller can also be designed by introducing the neural network algorithm to online model identification. Cheng et al. proposed a neural network-based controller to reduce the effect of complex nonlinearities between the end-effector and the driving object. The structure block diagram of the overall controller is shown in **Figure 11**. The control paradigm of piezoelectric stick-slip actuators is usually divided into two phases—the one-step control phase and the sub-step control phase. In the one-step control phase, when the error between the desired position and the actual position is less than the maximum single-step displacement length by continuous sawtooth wave

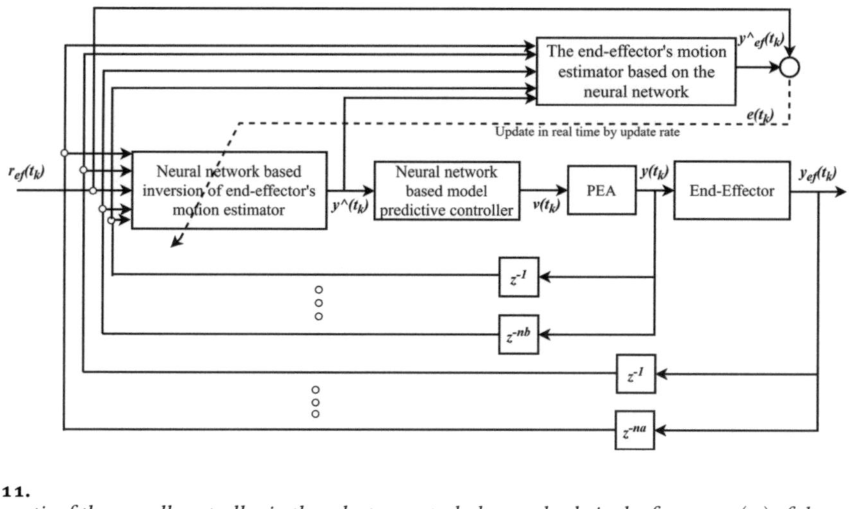

Figure 11.
The schematic of the overall controller in the sub-step control phase—the desired reference $r_{ef}(t_k)$ of the end-effector; the real displacement $y_{ef}(t_k)$ of the end-effector; the estimated displacement $y^{ef}(t_k)$ of the end-effector; the desired displacement $y^{(t_k)}$ of the driving object/PEA; the input voltage $v(t_k)$ applied to the PEA; the real displacement $y(t_k)$ of the driving object/PEA.

excitation, the controller switches to the sub-step control phase. In the experiment, the steady-state tracking error is kept within 50 nm, realizing ultra-precise motion control at the nanometer level [49].

Oubellil et al. applied proportional control to the macro motion control of nanorobots based on the piezoelectric stick-slip motion principle. Under macro motion control, the amplitude and frequency of the sawtooth wave voltage signal are adjusted by proportional control. When switching to scan control mode, Hammerstein dynamic model based on the PI hysteresis model is established, and then H∞ robust control scheme based on the model is designed. The hybrid stepper/scan controller can effectively meet the stability, robustness, hysteresis, and accuracy of multi-target nanorobots [50]. Oubellil et al. also applied piezoelectric stick-slip actuators to the nanorobot system of fast scanning probe microscope. To meet the requirements of fast scanning in closed-loop bandwidth and vibration reduction, the uncertain model of the piezoelectric actuator was defined by the multi-linear approximation method. A 2-DOF H∞ control scheme is designed to provide robust performance for the positioning of the nanorobot system. The fast and accurate positioning of the piezoelectric stick-slip actuator is realized [51].

In addition to its characteristics, the model of piezoelectric stick-slip actuators can also absorb the modeling mode of the piezoelectric stack. In a sense, due to the coupling relationship of the structure, the model can be regarded as an inclusion relation. The control mode of piezoelectric stick-slip actuators can also be the control mode of the piezoelectric stack, such as model-based feedforward control, inversion of control, sliding mode control method, active disturbance rejection control and some intelligent control methods can be applied to the precise control of piezoelectric stick-slip actuators. The research on piezoelectric stack also has reference significance in the precise control of piezoelectric stick-slip actuators.

Sliding mode controller often appears in the control of piezoelectric stack actuators nonlinear system [52, 53]. It is an effective and simple method to deal with the defects and uncertainties of a nonlinear system model. Sliding mode control is not dependent on an accurate mathematical model, which makes it popular in nonlinear system control of piezoelectric actuators. Li et al. proposed a sliding mode controller with disturbance estimation is designed for piezoelectric actuators. The Bouc-Wen model is chosen to describe the input and output relations of the piezoelectric actuator, and a particle swarm optimization algorithm is used for real-time identification of the model parameters. Considering the external and own uncertain disturbances, adaptive control rules are introduced to change the controller parameters. Experimental results show that the proposed controller can significantly improve the transient response speed of the system [54]. Mishra et al. designed a new continuous third-order sliding mode robust control scheme for the hinged piezoelectric actuator. To ensure the overall stability of the closed-loop system, a disturbance estimator was designed to counteract the effects of external disturbances and nonlinearities [55]. Xu Q et al. proposed an enhanced model predictive discrete sliding mode control (MPDSMC) with proportional-integral (PI) sliding mode function and a novel continuous third-order integral terminal sliding mode control (3-ITSMC) strategy [56, 57].

Because of the hysteresis nonlinearity of the piezoelectric actuator and the existence of system vibration and external disturbance, the robustness of the controller is usually required. Wei et al. proposed a variable bandwidth active disturbance rejection control method for piezoelectric actuators. The control method of the nanopositioning system is based on a cascade model of the hysteresis model and the system structure.

Information about all uncertainties and disturbances excluded items in the model is estimated by a time-varying extended state observer (TESO). Afterwards, a variable bandwidth controller based on the control error is designed. Its control system is shown in **Figure 12**. $z1$, $z2$, $z3$are the states of time-varying extended state observer, $b0$ is adjustable coefficient, and d is a disturbance. A series of experiments show that the proposed controller has a higher response speed and stronger anti-interference ability than the traditional active disturbance rejection controller [58].

Neural networks are widely used in the design of adaptive controllers for nonlinear systems because of their strong self-learning ability. In view of the system uncertainty and hysteresis nonlinearity of piezoelectric actuator, Li et al. proposed a neural network self-tuning control method. Two nonlinear function variables about hysteresis output are established and two neural networks are introduced to identify the two hysteresis function variables on line, respectively. Experiments verify that the neural network self-tuning controller has a good track tracking effect [59]. Napole et al. proposed a new method combining super torsion algorithm (STA) and artificial neural network (ANN) to improve the tracking accuracy of high voltage stack actuator [60]. Lin et al. proposed a dynamic Petri fuzzy cerebellar (DPFC) model joint controller for magnetic levitation system (MLS) and two-axis piezoelectric ceramic motor (LPCM) drive system, which is used to control the position of MLS metal ball and track tracking of the two-axis LPCM drive system. The experimental results also show that this method can obtain a high-precision trajectory tracking response [61].

The neural network has a strong self-learning ability and can approach complex nonlinear functions. It plays an important role in the design of the piezoelectric stick-slip actuators controller. In addition to neural network and sliding mode control, data-driven model-free adaptive control is also suitable for systems with model uncertainty. Model-free adaptive control (MFAC) as a typical data-driven control method, this method was proposed in Mr. Hou Z's doctoral thesis in 1994 [62]. In the past two decades, both the continuous development and improvement of theoretical achievements, and the successful practical application in the fields of motor control, chemical industry, machinery and so on, have made MFAC become a new control theory with a systematic and rigorous framework. As for the application of model-free control in the piezoelectric stack, Muhammad designed a data-driven feedforward controller and feedback controller. To avoid chattering caused by noise and affect the convergence of the learning process, several rules about parameters are also proposed. The experimental results show that the controller can realize high-precision position tracking at low frequency [63].

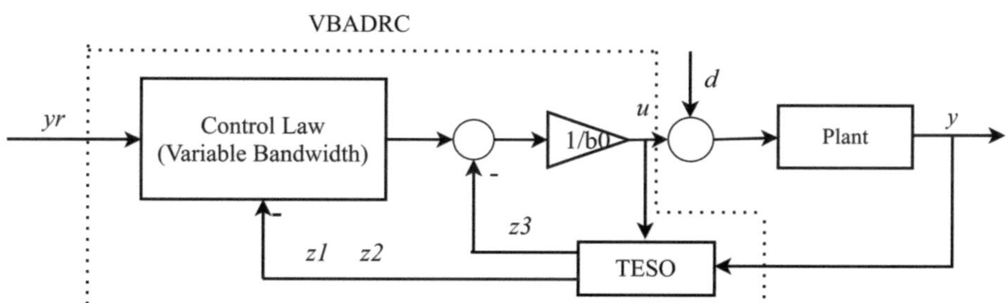

Figure 12.
The variable bandwidth active disturbance rejection control.

4. Conclusions

Piezoelectric stick-slip actuators have great potential in the field of precision operation. However, whether at the experimental level or in the application, the hysteresis nonlinearity of the piezoelectric stick-slip actuator, the complex friction motion relationship in the driving mechanism, its vibration and external disturbance will have a great impact on its motion control accuracy, so that the piezoelectric stick-slip actuator cannot achieve the ideal output performance. In this paper, the modeling and control of piezoelectric stick-slip actuators are summarized and studied.

In the aspect of modeling, the existing mathematical models describing the hysteresis characteristics of piezoelectric stick-slip actuators and the mathematical models of complex friction relationships in the structure are introduced. Hysteresis models mainly include Prandtl-Ishlinskii (PI) model, Krasnosel'skii-Pokrovskii (KP) model, Preisach model, Bouc-Wen model, and Rayleigh model. In terms of the friction model, the existing dynamic friction and static friction models are mainly introduced. The model of piezoelectric stick-slip actuators usually includes the hysteresis model and friction model. In the modeling part, the mathematical model of piezoelectric stick-slip actuators proposed by people is summarized and studied, which provides a reference for the control and model analysis of piezoelectric stick-slip actuators.

In terms of control, according to open-loop control and closed-loop control, this paper summarizes and studies the efforts made by people to make up for control accuracy, and summarizes many control cases, such as feedforward control, sliding mode control, PID control, neural network control, and so on. In the future development of piezoelectric stick-slip actuators, opportunities and difficulties coexist. The control mode can effectively make up for the output performance of piezoelectric stick-slip actuators and make them meet the actual need in various complex environments.

Based on this paper, a more comprehensive dynamics model can be developed in the future by analyzing the characteristics of piezoelectric viscous-slip actuators in-depth to extend to actuators of different mechanical structures. By combining various control methods to eliminate system nonlinearity, higher accuracy and precision motion can be achieved. With the combination of intelligent control field and piezoelectric actuators, piezoelectric stick-slip actuators will be applied to more fields in the future.

Acknowledgements

This work was financially supported by the Science and Technology Development Plan of Jilin Province (Nos. 20200201057JC and 20190201108JC), and the Technology Research Planning Project of Education Department of Jilin Province (No. JJKH20220690KJ).

Conflict of interest

The author(s) declared no potential conflicts of interest with respect to the research, authorship, and/or publication of this article.

Author details

Zhenguo Zhang[1], Piao Fan[2], Yikun Dong[2], Shuai Yu[2], Keping Liu[2]* and Xiaohui Lu[1]

1 School of Mechatronic Engineering, Changchun University of Technology, Changchun, China

2 School of Electrical and Electronic Engineering, Changchun University of Technology, Changchun, China

*Address all correspondence to: liukeping@ccut.edu.cn

References

[1] Wei J, Qiu Z, Han J, Wang Y. Experimental comparison research on active vibration control for flexible piezoelectric manipulator using fuzzy controller. Journal of Intelligent and Robotic Systems. 2010;**59**:31-56. DOI: 10.1007/s10846-009-9390-2

[2] Croft D, Shed G, Devasia S. Creep, hysteresis, and vibration compensation for piezoactuators: Atomic force microscopy application. Journal of Dynamic Systems Measurement and Control. 2001;**123**:35. DOI: 10.1115/1.1341197

[3] Martin PM, Matson DW, Bennett WD, Lin Y, Hammerstrom DJ. Laminated plastic microfluidic components for biological and chemical systems. Journal of Vacuum Science & Technology A: Vacuum, Surfaces, and Films. 1999;**17**:2264-2269. DOI: 10.1116/1.581758

[4] Matthias H. Piezoelectric inertia motors—A critical review of history, concepts, design, applications, and perspectives. Actuators. 2017;**6**:7-7. DOI: 10.3390/act6010007

[5] Peng J, Chen X. Modeling of piezoelectric-drive stick-slip actuator. IEEE/ASME Transactions on Mechatronics. 2011;**16**:394-399. DOI: 10.1109/tmech.2010.2043849

[6] Song G, Zhao J, Zhou X, DeAbreu-Garcia J. Tracking control of a piezoceramic actuator with hysteresis compensation using inverse Preisach model. IEEE/ASME Transactions on Mechatronics. 2005;**10**:198-209. DOI: 10.1109/TMECH.2005.844708

[7] Li J, Zhou X, Zhao H, Shao M, Li N, Zhang S, et al. Development of a novel parasitic-type piezoelectric actuator. IEEE/ASME Transactions on Mechatronics. 2017;**22**:541-550. DOI: 10.1109/tmech.2016.2604242

[8] Adriaens HJMTS, De Koning WL, Banning R. Modeling piezoelectric actuators. IEEE/ASME Transactions on Mechatronics. 2000;**5**:331-341. DOI: 10.1109/3516.891044

[9] Jiles DC, Atherton DL. Theory of ferromagnetic hysteresis. Journal of Magnetism and Magnetic Materials. 1986;**61**:48-60. DOI: 10.1016/0304-8853(86)90066-1

[10] Ikuta K, Tsukamoto M, Hirose S. Mathematical model and experimental verification of shape memory alloy for designing micro actuator. IEEE Micro Electro Mechanical Systems. 1991;**1991**:103-108. DOI: 10.1109/MEMSYS.1991.114778

[11] Preisach F. Über die magnetische Nachwirkung. Zeitschrift für Physik a Hadrons and Nuclei. 1935;**94**:277-302. DOI: 10.1007/BF01349418

[12] Li C, Tan Y. A neural networks model for hysteresis nonlinearity. Sensors & Actuators A Physical. 2004;**112**:49-54. DOI: 10.1016/j.sna.2003.11.016

[13] Li C, Tan Y. Modelling Preisach-type hysteresis nonlinearity using neural network. International Journal of Modelling and Simulation. 2007;**27**:233-241. DOI: 10.2316/Journal.205.2007.3.205-4375

[14] Macki J, Nistri P, Zecca P. Mathematical models for hysteresis. SIAM Review. 1993;**35**:94-123. DOI: 10.1137/1035005

[15] Krasnoselskii M, Pokrovskii A. Systems with Hysteresis. Moscow:

Springer; 1989. pp. 43-66. DOI: 10.1007/978-3-642-61302-9

[16] Coleman B, Hodgdon M. On a class of constitutive relations for ferromagnetic hysteresis. Archive for Rational Mechanics and analysis. 1987; **99**:375-396. DOI: 10.1007/BF00282052

[17] Su CY, Stepanenko Y, Svoboda J, et al. Robust adaptive control of a class of nonlinear systems with unknown backlash-like hysteresis. IEEE Transactions on Automatic Control. 2000;**45**:2427-2432. DOI: 10.1109/9.895588

[18] Bouc R. Forced vibration of mechanical systems with hysteresis. 1967

[19] Wen Y. Method for random vibration of hysteretic systems. Journal of the engineering mechanics division. 1976;**102**:249-263. DOI: 10.1061/JMCEA3.0001360

[20] Ikhouane F, MañOsa V, Rodellar J. Adaptive control of a hysteretic structural system. Automatica. 2005;**41**: 225-231. DOI: 10.1016/j.automatica.2004.08.018

[21] Ikhouane F, Rodellar J. On the Hysteretic Bouc–Wen Model. Nonlinear Dynamics. 2005;**42**:79-95. DOI: 10.1007/s11071-005-0070-x

[22] Gan J, Zhang X, Wu H. Tracking control of piezoelectric actuators using a polynomial-based hysteresis model. AIP Advances. 2016;**6**:065204-065210. DOI: 10.1063/1.4953597

[23] Cheng L, Liu W, Hou ZG, et al. Neural-network-based nonlinear model predictive control for piezoelectric ceramic actuators. IEEE Transactions on Industrial Electronics. 2015;**62**:7717-7727. DOI: 10.1109/TIE.2015.2455026

[24] Zhang M, Damjanovic D. Quasi-rayleigh model for modeling hysteresis of piezoelectric actuators. Smart Materials and Structures. 2020;**29**:1-16. DOI: 10.1088/1361-665X/ab874b

[25] Li P, Zhang D, Hu J, et al. Hysteresis modelling and feedforward control of piezoelectric actuator based on simplified interval type-2 fuzzy system. Sensors. 2020;**20**:2587. DOI: 10.3390/s20092587

[26] Awrejcewicz J, Olejnik P. Analysis of dynamic systems with various friction laws. Applied Mechanics Reviews. 2005; **58**:389–411. DOI: 10.1115/1.2048687

[27] Zhang Z. Modeling and understanding of directional friction on a fully lubricated surface with regular anisotropic asperities. Master of engineering thesis. University of Saskatchewan. 2010. DOI: 10.1007/s11012-010-9303-2

[28] Li CB, Pavelescu D. The friction-speed relation and its influence on the critical velocity of stick-slip motion. Wear. 1982;**82**:277-289. DOI: 10.1016/0043-1648(82)90223-X

[29] Dahl PR. Solid friction damping of mechanical vibrations. AIAA Journal. 1976;**14**:1675-1682. DOI: 10.2514/3.61511

[30] Canudas W, Olsson H, Astrom KJ, Lischinsky P. A new model for control of systems with friction. IEEE Transactions on Automatic Control. 1995;**40**:419-425. DOI: 10.1109/9.376053

[31] Swevers J, l-Bender F, Ganseman C Projogo T. An integrated friction model structure with improved presliding behavior for accurate friction compensation. IEEETransaction on Automatic Control. 2000;**45**:675-686. DOI: 10.1109/9.847103

[32] Wang Y, Xu M, Shao S, et al. A Novel Stick-Slip Type Rotary Piezoelectric Actuator. Advances in Materials Science and Engineering. 2020;**2020**:1-11. DOI: 10.1155/2020/2659475

[33] Nguyen H, Teidelt E, Popov V, et al. Modeling and waveform optimization of stick–slip micro-drives using the method of dimensionality reduction. Archive of Applied Mechanics. 2014;**86**:1-15. DOI: 10.1007/s00419-014-0934-y

[34] Shao Y, Xu M, Shao S, et al. Effective dynamical model for piezoelectric stick–slip actuators in bi-directional motion. Mechanical Systems and Signal Processing. 2020;**145**:106964. DOI: 10.1016/j.ymssp.2020.106964

[35] Wang X, Zhu L, Huang H. A dynamic model of stick-slip piezoelectric actuators considering the deformation of overall system. IEEE Transactions on Industrial Electronics. 2021;**68**:11266-11275. DOI: 10.1109/TIE.2020.3032922

[36] Holub O, Špiller M, Hurák Z, editors. Stick-slip based micropositioning stage for transmission electron microscope. In: Proceedings of the IEEE International Workshop on Advanced Motion Control; 27-29 March 2006; Istanbul, Turkey: IEEE; 2006. p. 484–487

[37] Chang S, Du B. A precision piezodriven micropositioner mechanism with large travel range. Review of Entific Instruments. 1998;**69**:1785-1791. DOI: 10.1063/1.1148842

[38] Bergander A, Breguet J, Schmitt C, et al. Micropositioners for microscopy applications based on the stick-slip effect. International Symposium on Micromechatronics & Human Science. IEEE. 2000;**24**:213-216. DOI: 10.1109/MHS.2000.903315

[39] Spiller M, Hurak Z. Hybrid charge control for stick–slip piezoelectric actuators. Mechatronics. 2011;**21**:100-108. DOI: 10.1016/j.mechatronics.2010.09.002

[40] Chen J, Peng G, Hu H, et al. Dynamic hysteresis model and control methodology for force output using piezoelectric actuator driving. IEEE Access. 2020;**8**:205136-205147. DOI: 10.1109/ACCESS.2020.3037216

[41] Ha J, Fung R, Yang C. Hysteresis identification and dynamic responses of the impact drive mechanism. Journal of Sound & Vibration. 2005;**283**:943-956. DOI: 10.1016/j.jsv.2004.05.032

[42] Wei T, Khosla P, Riviere C. Feedforward controller with inverse rate-dependent model for piezoelectric actuators in trajectory-tracking applications. IEEE/ASME Transactions on Mechatronics. 2007;**12**:134-142. DOI: 10.1109/TMECH.2007.892824

[43] Zhang M, Liu Z, Zhu Y. Inverse rate-dependent rayleigh model based feedforward control for piezoelectric-driven mechanism. IEEE Access. 2020;**8**:194808-194819. DOI: 10.1109/ACCESS.2020.3033845

[44] Zhong B, Jin Z, Zhu J, et al. Double closed-loop control of a trans-scale precision positioning stage based on the inertial stick-slip driving. Sensors and Actuators A: Physical. 2019;**297**:111547. DOI: 10.1016/j.sna.2019.111547

[45] Rong W, Liang S, Wang L, et al. Model and control of a compact long-travel accurate-manipulation platform. IEEE/ASME Transactions on Mechatronics. 2017;**22**:402-411. DOI: 10.1109/TMECH.2016.2597168

[46] Rakotondrabe M, Haddab Y, Lutz P. Development, modeling, and control of a

micro-/nanopositioning 2-DOF stick-slip device. IEEE/ASME Transactions on Mechatronics. 2010;**14**:733-745. DOI: 10.1109/TMECH.2009.2011134

[47] Theik C, Mazlan A. Active vibration control of an inertia-type piezoelectric actuator based suspended handle using PID-AFC controller. In: IEEE Symposium on Industrial Electronics & Applications (ISIEA). 2020. DOI: 10.1109/ISIEA49364.2020.9188127

[48] Cao Y, Chen X. An ARX-based PID-sliding mode control on velocity tracking control of a stick-slip piezoelectric-driven actuator. Modern Mechanical Engineering. 2015;**5**:10-19. DOI: 10.4236/mme.2015.51002

[49] Cheng L, Liu W, Yang C, Huang T, et al. A neural-network-based controller for piezoelectric-actuated stick-slip devices. IEEE Transactions on Industrial Electronics. 2018;**65**:2598-2607. DOI: 10.1109/TIE.2017.2740826

[50] Oubellil R, Voda A, Boudaoud M, et al. Mixed stepping/scanning mode control of stick-slip SEM-integrated nano-robotic systems. Sensors & Actuators A Physical. 2018;**285**:258-268. DOI: 10.1016/j.sna.2018.08.042

[51] Oubellil R, Voda A, Boudaoud M, et al. Robust control strategies of stick-slip type actuators for fast and accurate nanopositioning operations in scanning mode. Proceedings of the Mediterranean Conference on Control and Automation (MED); 16-19 June 2015; Torremolinos, Spain: IEEE; 2015. p. 650-655. DOI: 10.1109/MED.2015.7158820

[52] Qi N, Zhang C, Yuan J. Observer based sliding mode control for subsonic piezocomposite plate involving time varying measurement delay. Measurement and Control London: Institute of Measurement and Control; 2021;**54**:1-10. DOI: 10.1177/0020294020983373

[53] Yu S, Xie M, Ma J, et al. Precise robust motion tracking of a piezoactuated micropuncture mechanism with sliding mode control. Journal of the Franklin Institute. 2021; **358**:4410-4434. DOI: 10.1016/j.jfranklin.2021.04.025

[54] Li Y, Xu Q. Adaptive sliding mode control with perturbation estimation and PID sliding surface for motion tracking of a piezo-driven micromanipulator. IEEE Transactions on Control Systems Technology. 2010;**18**:798-810. DOI: 10.1109/TCST.2009.2028878

[55] Mishra J, Xu Q, Yu X, et al. Precision position tracking for piezoelectric-driven motion system using continuous third-order sliding mode control. IEEE/ASME. 2018;**23**:1521-1531. DOI: 10.1109/TMECH.2018.2853737

[56] Xu Q, Li Y. Model predictive discrete-time sliding mode control of a nanopositioning piezostage without modeling hysteresis. Control Systems Technology, IEEE Transactions on. 2012; **20**:983-994. DOI: 10.1109/TCST.2011.2157345

[57] Xu Q. Continuous integral terminal third-order sliding mode motion control for piezoelectric nanopositioning system. IEEE/ASME Transactions on Mechatronics. 2017;**22**:1828-1838. DOI: 10.1109/TMECH.2017.2701417

[58] Wei W, Xia P, Xue W, et al. On the disturbance rejection of a piezoelectric driven nanopositioning system. IEEE Access. 2020;**8**:74771-74781

[59] Li W, Zhang C, Gao W, et al. Neural network self-tuning control for a piezoelectric actuator. Sensors (Basel, Switzerland). 2020;**20**:3342. DOI: 10.3390/s20123342

[60] Napole C, Barambones O, Derbeli M, et al. High-performance tracking for piezoelectric actuators using super-twisting algorithm based on A Review of Modeling and Control of Piezoelectric Stick-Slip Actuators artificial neural networks. Mathematics, MDPI. 2021;**9**:1-20. DOI: 10.3390/math9030244

[61] Lin C, Li H. Dynamic Petri fuzzy cerebellar model articulation controller design for a magnetic levitation system and a two-axis linear piezoelectric ceramic motor drive system. IEEE Transactions on Control Systems Technology. 2015;**23**:693-699. DOI: 10.1109/TCST.2014.2325897

[62] Hou Z, Jin S. Model free adaptive control: Theory and applications. 1st ed. Boca Raton: CRC Press; 2013. p. 399. DOI: 10.1201/b15752

[63] Muhammad S, Ashraf S, Mostefa M. Model-free data driven control for trajectory tracking of an amplified piezoelectric actuator. Sensors & Actuators A Physical. 2018;**279**:27-35. DOI: 10.1016/j.sna.2018.05.010